──────Statistics in Research and Development──────

To Pamela Joan

STATISTICS IN RESEARCH AND DEVELOPMENT

Roland Caulcutt
Statistics for Industry (UK) Ltd

LONDON   NEW YORK
CHAPMAN AND HALL

First published 1983 by
Chapman and Hall Ltd
11 New Fetter Lane, London EC4P 4EE
Published in the USA by Chapman and Hall
733 Third Avenue, New York NY10017

©1983 R. Caulcutt

Printed in Great Britain by Richard Clay
(The Chaucer Press) Ltd, Bungay, Suffolk

Typeset in Great Britain by Keyset Composition
Colchester, Essex

ISBN 0 412 23720 2

British Library Cataloguing in Publication Data

Caulcutt, R.

  Statistics in research and development.
  1. Mathematical statistics
  I. Title
  519.5   QA276

ISBN 0-412-23720-2

Library of Congress Cataloging in Publication Data

Caulcutt, R.

  Statistics in research and development.
  Bibliography: p.
  Includes index.
  1. Statistics.   I. Title.
  QA276.12.C387 1982   001.4'22   82-12800

ISBN 0-412-23720-2

——Contents——

————Statistical Tables————

———Preface———

This book is based upon material originally prepared for courses run by Statistics for Industry (UK) Ltd. Over a period of some four or five years the notes were repeatedly extended and refined as the author developed a clearer image of the needs of the rather heterogeneous course members. It is hoped that this effort will not have been in vain and that this final version will be of use to engineers, physicists and biologists, in addition to the chemists for whom the original notes were intended.

The approach adopted throughout this book has two distinctive features, which also characterize the statistical courses on which the book is based. This approach is both 'problem centred' and 'non-mathematical', thus enabling the reader to concentrate on three essential elements:

(a) how to use statistical techniques,
(b) how to interpret the results of statistical analysis,
(c) how to check the assumptions underlying the statistical techniques.

The spread of microcomputers into laboratories and the increasing availability of statistical computer programs has created the possibility of scientists and technologists carrying out their own statistical analysis, without reference to a statistician. It is hoped that this book will be very helpful to any scientist who attempts to 'go it alone', but the book will certainly *not* convert a scientist or technologist into a statistician. In fact the reader would be well advised to regard this as an 'introductory' book, despite the great statistical heights reached in certain chapters of Part Two. Several of the texts listed in the bibliography could be used to consolidate and to broaden any understanding gained from this book.

Many of the statisticians and scientists who lecture on Statistics for Industry courses have made numerous suggestions during the preparation of this book. I am particularly grateful to Dick Boddy, Derrick Chamberlain, John Sykes and David Arthur for their advice and support. Dr A. F. Bissell offered many constructive comments on an earlier draft of this book, for which I am very grateful. This policy of continuous improvement has, hopefully, resulted in a more readable book. It has certainly made great demands on my typist, Christine Robinson, who has managed to produce a beautiful manuscript from a collection of mis-spelt and unpunctuated jottings. This does not absolve myself from the responsibility for any errors that remain.

Statistics for Industry (UK) Ltd run many courses in applied statistics at a variety of venues throughout the year. The courses include:

Introduction and significance testing
Statistics in research and development
Statistics for analytical chemists
Design of experiments
Statistical quality control

New courses are introduced every year and full details of all courses can be obtained from:

The Conference Secretary,
Statistics for Industry (UK) Ltd,
14 Kirkgate,
Knaresborough,
N. Yorkshire,
HG5 8AD.
Tel: (0423) 865955

PART ONE

1

What is statistics?

The purpose of this chapter is to single out the prime elements of statistics so that you will not lose sight of what is most important as you work through the detail of subsequent chapters. Certain words will be introduced which will recur throughout the book; they are written in bold letters in this chapter and each one is underlined when it first appears. Let us start with a simple definition:

> Statistics is a body of knowledge which can be of use to anyone who has taken a **sample**.

It is appropriate that the first word which has been presented in bold letters is '**sample**', since its importance cannot be overemphasized. Many people in the chemical and allied industries occasionally find themselves in the unenviable position of needing to investigate a **population** but are constrained by the available resources to examine only a **sample** taken from the **population**. In essence a **population** is simply a large group. In some situations the **population** is a large group of people, in other situations it is a large group of inanimate objects, though it can in some cases be more abstract. Let us consider some examples of situations in which someone has taken a **sample**.

Example 1.1
A works manager wishes to **estimate** the average number of days absence due to sickness for his employees during 1977. From the files which contain details of absences of the 2300 employees his secretary selects the record cards of 50 employees. She calculates that the average number of days of sickness absence of the 50 employees during 1977 is 3.61 days.
■■■

In this example the **population** consists of 2300 employees whilst the **sample** consists of 50 employees. After the **sample** was taken there was no doubt about the *average sickness absence* of the employees in the **sample**. It is the **population** average about which we are uncertain. The works manager intends to use the **sample** average (3.61 days) as an **estimate** of the **population** average. Common sense would suggest that there are dangers to be heeded. Perhaps it is worthwhile to spell out several points, some of which may be obvious:

(a) If the secretary had selected a different set of 50 employees to include in the **sample** the sample average would almost certainly have had a different value.
(b) The **sample** average (3.61) is very unlikely to be equal to the **population** average. (More formally we could say that the **probability** of the **sample** average being equal to the **population** average is very small.)
(c) Whether or not the **sample** average is close to the **population** average will depend on whether or not the **sample** is **representative** of the **population**.
(d) No one can guarantee that a **sample** will be **representative** of the **population** from which it came. We can only hope that, by following a reputable procedure for taking the **sample**, we will end up with a **sample** which is **representative** of the **population**. It is unfortunately true, as Confucius may have pointed out, that 'He who takes a **sample** takes a risk'.
(e) One reputable procedure for taking a **sample** is known as **random sampling**. The essential feature of this method is that every member of the **population** has the same chance (or **probability**) of being included in the **sample**. The end product of **random sampling** is known as a **random sample**, and all the statistical techniques which are introduced in this book are based upon the assumption that a **random sample** has been taken.

Example 1.2
Nicoprone is manufactured by a batch production process. The plant manager is worried about the percentage impurity, which appears to be higher in recent batches than it was in batches produced some months ago. He suspects that the impurity of the final product may depend on the presence of polystyline in the resin which is one of the raw materials of the process. The supplier of the resin has agreed that the polystyline content will not exceed 1% on average and that no single bag will contain more than 2.5%. Approximately 900 bags of resin are in the warehouse at the present time. The warehouse manager takes a **sample** of 18 bags by selecting every 50th bag on the shelf. From each of the selected bags a 20 gram **sample*** of resin is taken. The determinations of polystyline content are:

1.6% 0.5% 3.1% 0.7% 0.8% 1.7% 1.4% 0.8% 1.1%
0.9% 2.4% 0.6% 2.2% 2.9% 0.3% 0.5% 1.0% 1.3%

We can easily calculate that the average polystyline content is equal to 1.32% and we notice that two of the determinations exceed 2.5%. So the **sample** average is certainly greater than the specified limit of 1.0%, but we need to consider what this average *might* have been if the warehouse manager had selected a different sample of bags, or perhaps we should ask ourselves what the average polystyline content would have been if he had sampled *all* the bags in the warehouse.

■■■

*Note the different usage of the word **sample** by the chemist and the statistician. In this example the chemist speaks of *18 samples* whilst the statistician speaks of *one sample* containing 18 items.

Questions like these will be answered in subsequent chapters. At this point we will probe the questions more deeply by translating into the language of the statistician. In this more abstract language we would ask 'What conclusions can we draw concerning the **population**, based upon the **sample** that has been examined?' This in turn prompts two further, very important questions, 'What exactly is the **population** about which we wish to draw conclusions?' and perhaps surprisingly 'What exactly is the **sample** on which the conclusions will be based?'

It is easier to answer these questions in reverse order. The **sample** can be looked upon as either:

(a) 18 bags of resin,
(b) 18 quantities of resin, each containing 20 g, or
(c) 18 measurements of polystyline content.

Whether it is better to take (a), (b) or (c) will depend upon such chemical/ physical considerations as the dispersion of the polystyline within the resin and how the **variability** between bags compares with the **variability** within bags. Let us assume that each measurement gives a true indication of the polystyline content of the bag it represents and we will take (a) as our definition of the **sample**.

If our **sample** consists of 18 bags of resin then our **population** must also consist of bags of resin. The 18 bags were chosen from those in the warehouse, so it might seem reasonable to define our **population** as 'the 900 bags of resin in the warehouse', but does this cover *all* the resin about which the manager wishes to draw conclusions? He may wish to define the **population** in such a way as to include all bags of resin received in the past from this supplier and all bags to be received in the future. Before taking this bold step he would need to ask himself, 'Is the **sample** I have taken **representative** of the **population** I wish to define?'

It would obviously not be possible to take a **random sample** from such a **population** since the batches to be received in the future do not yet exist. Whenever we attempt to predict the future from our knowledge of the past we are talking about a **population** from which we cannot take a **random sample**. We may, nonetheless, be confident that our **sample** is **representative**. (The statistician prefers to discuss **random samples** rather than **representative samples** since the former are easier to define and are amenable to the tools of **probability** theory. The statistician does not, however, wish to see the statistical tail wagging the scientific/technical dog.)

Even if the plant manager defines his **population** as 'the 900 bags of resin in the warehouse' he still hasn't got a **random sample**. When the warehouse manager selected every 50th bag from the shelf he was practising what is known as **systematic sampling**. This is a procedure which is often used in the inspection of manufactured products and there is a good chance that **systematic sampling** will give a **representative sample** provided there are no hidden patterns in the **population**.

It has already been stated that the statistical techniques in this book are built upon the mathematical basis of **random sampling**, but this is only *one* of the many assumptions used by statisticians. An awareness of these assumptions is just as important to the scientist or technologist as the ability to select the appropriate statistical technique. For this reason a substantial part of Chapter 7 is devoted to the assumptions underlying the important techniques presented in Chapters 4 to 6.

———2———
Describing the sample

2.1   Introduction

In the previous chapter we focused attention on three words which are very important in statistics:

Sample
Population
Variability.

When a scientist or technologist is using statistical techniques he is probably attempting to make a generalization, based upon what he has found in one or more samples. In arguing from the particular to the general he will also be inferring from the *sample* to the *population*. Whilst doing so it is essential that he takes account of the *variability* within the sample(s).

It is the variability in the sample that alerts the scientist to the presence of random variation. Only by taking account of this random variation can we have an objective procedure for distinguishing between real and chance effects. Thus a prerequisite of using many statistical techniques is that we should be able to *measure* or describe the variability in a set of data.

In this chapter we will examine simple methods of describing variability and in doing so we will confine our attention to the *sample*.

2.2   Variability in plant performance

Higson Industrial Chemicals manufacture a range of pigments for use in the textile industry. One particular pigment, digozo blue, is made by a well established process on a plant which has recently been renovated. During the renovation various modifications were incorporated, one of which made the agitation system fully automatic. Though this program of work was very successful in reducing the number of operators needed to run the plant, production of digozo blue has not been completely trouble free since the work was completed. Firstly, the anticipated increase in yield does not appear to have materialized, and secondly, several batches have been found to contain a disturbingly large percentage of a particular impurity.

The plant manager has suggested that the plant is unable to perform as well as expected because the agitator cannot cope with the increased capacity of the vessel. With the new control system the agitation speed is automatically reduced for a period of two minutes whenever the agitator drive becomes overloaded. The plant manager is of the opinion that it is during these slow periods that the reaction is unsatisfactory.

Whether his diagnosis is correct or not there can be no doubt that these overloads do occur because each incident is automatically recorded on a data logger. The number of overloads which occurred has been tabulated by the chief chemist (production) in Table 2.1 for the first 50 batches produced since the renovation was completed. Also tabulated is the yield and the percentage impurity of each batch.

Table 2.1   Yield, impurity and overloads in 50 batches of digozo blue pigment

Batch	Yield	Impurity	Number of overloads	Batch	Yield	Impurity	Number of overloads
1	69.0	1.63	0	26	69.5	1.78	1
2	71.2	5.64	4	27	72.6	2.34	1
3	74.2	1.03	2	28	66.9	0.83	0
4	68.1	0.56	2	29	74.2	1.26	0
5	72.1	1.66	1	30	72.9	1.78	1
6	64.3	1.90	4	31	69.6	4.92	3
7	71.2	7.24	0	32	76.2	1.58	0
8	71.0	1.62	1	33	70.4	5.13	5
9	74.0	2.64	2	34	68.3	3.02	1
10	72.4	2.10	2	35	65.9	0.19	3
11	67.6	0.42	0	36	71.5	2.00	1
12	76.7	1.07	1	37	69.9	1.15	0
13	69.0	1.52	1	38	70.4	0.66	0
14	70.8	11.31	2	39	74.1	3.24	3
15	78.0	2.19	0	40	73.5	2.24	2
16	73.6	3.63	2	41	68.3	2.51	2
17	67.3	1.07	1	42	70.1	2.82	1
18	72.5	3.89	0	43	75.0	0.37	0
19	70.7	2.19	2	44	78.1	2.63	1
20	69.3	0.71	3	45	67.2	4.17	2
21	75.8	0.83	1	46	71.5	1.38	0
22	63.8	1.38	0	47	73.0	3.63	4
23	72.6	2.45	3	48	71.8	0.76	2
24	70.5	2.34	0	49	68.8	3.09	3
25	76.0	8.51	5	50	70.1	1.29	0

The chief chemist hopes that the data in Table 2.1 will help him to answer several questions, including:

(a) If the present situation is allowed to continue what percentage of batches will contain more than 5% impurity?

(b) If the present situation is allowed to continue what percentage of batches will give a yield of less than 66%?

(c) Is the quality or the yield of a batch related to the number of overloads that occurred during the manufacture of the batch?

When we attempt to answer questions like those above, we make use of statistical techniques which come under the heading of *inferential statistics*. To answer the questions we would need to *infer* the condition of future batches from what we have found when examining the 50 batches which constitute our sample. We must postpone the use of inferential statistics until we have explored much simpler techniques which come under the alternative heading of *descriptive statistics*. The use of descriptive statistics will help us to summarize the data in Table 2.1 as a prelude to using it for inferential purposes.

2.3 Frequency distributions

The yield values in Table 2.1 can be presented in a compact form if they are re-tabulated as in Table 2.2.

Table 2.2   A frequency distribution for the yield of 50 batches

Yield (%)	62.0 to 63.9	64.0 to 65.9	66.0 to 67.9	68.0 to 69.9	70.0 to 71.9	72.0 to 73.9	74.0 to 75.9	76.0 to 77.9	78.0 to 79.9	Total
Number of batches	1	2	4	10	13	9	6	3	2	50

To produce Table 2.2 each of the 50 batches has been allocated to one of the nine groups listed in the upper row of the table. The number of batches in each group is indicated by the number below. These numbers in the bottom row of the table are known as *frequencies* and the whole table is often referred to as a frequency distribution. The frequencies do, of course, add up to 50 as there are 50 batches in the sample.

In Table 2.2 then we can see at a glance exactly how many of the 50 batches have a yield between 66.0 and 67.9, say. By comparing the frequencies for the different groups we can easily see that many batches have been allocated to the three groups in the centre of the table with very few batches in the end groups.

For some purposes it is convenient to express frequencies as percentages of the total frequency or as proportions of the total frequency. The percentage

Table 2.3   Percentage frequency distribution – yield

Yield (%)	62.0 to 63.9	64.0 to 65.9	66.0 to 67.9	68.0 to 69.9	70.0 to 71.9	72.0 to 73.9	74.0 to 75.9	76.0 to 77.9	78.0 to 79.9	Total
Percentage of batches	2	4	8	20	26	18	12	6	4	100

Table 2.4   Proportional frequency distribution – yield

Yield (%)	62.0 to 63.9	64.0 to 65.9	66.0 to 67.9	68.0 to 69.9	70.0 to 71.9	72.0 to 73.9	74.0 to 75.9	76.0 to 77.9	78.0 to 79.9	Total
Proportion of batches	0.02	0.04	0.08	0.20	0.26	0.18	0.12	0.06	0.04	1.00

frequencies will add up to 100%, of course, and the proportions will add up to 1.0, as we see in Table 2.3 and 2.4.

We could also produce frequency distributions for the other two columns of data in Table 2.1, as in Tables 2.5 and 2.6. There is no necessity to group the data for 'number of overloads' as there are only six different values recorded for this variable.

Table 2.5   Frequency distribution – number of overloads

Number of overloads	0	1	2	3	4	5	Total
Number of batches	15	13	11	6	3	2	50

Table 2.6   Frequency distribution – percentage impurity

Percentage impurity	Number of batches	Percentage impurity	Number of batches
0.00–0.99	9	6.00–6.99	0
1.00–1.99	16	7.00–7.99	1
2.00–2.99	12	8.00–8.99	1
3.00–3.99	6	9.00–9.99	0
4.00–4.99	2	10.00–10.99	0
5.00–5.99	2	11.00–11.99	1
		Total	50

The histogram and the bar chart

The frequency distribution is a simple tabulation which gives us an overall appreciation of a set of data. It is particularly useful if the data set is large. On the other hand, it could be argued that we only get the maximum benefit from a frequency distribution if we display it in a *pictorial* form. Figs 2.1, 2.2 and 2.4 represent the frequency distributions of our three variables, yield, impurity and number of overloads.

Fig 2.1 and Fig. 2.2 are known as histograms. In both diagrams the height of a bar (or block) tells us the number of batches which have yield (or impurity) in a particular range. We note that Fig. 2.1 and Fig. 2.2 are very different in shape, with Fig. 2.1 being virtually *symmetrical* whilst Fig. 2.2 is very *skewed*. Perhaps

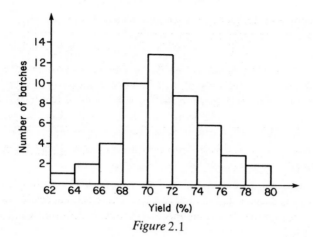

Figure 2.1

we should have expected different shapes for the distributions of these two variables. In Fig. 2.1 we see that a large number of batches are included in the middle three blocks, with yield between 68 and 74, and there are just a few batches in the tails of the distribution. In Fig. 2.2 we see that most of the batches are included in the three blocks to the left of the picture, with impurity between 0% and 3%. It would not be possible for this distribution to 'tail-off' in *both* directions since we are already up against the lower limit (0%).

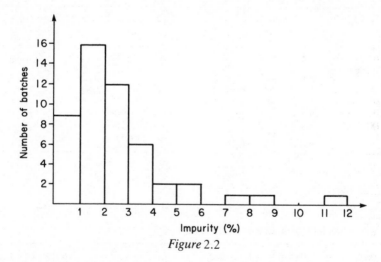

Figure 2.2

Whatever the *reasons* for the yield and the impurity of our 50 batches having very different frequency distributions, how does the difference in shape affect our confidence in *future* batches? It is probably fair to say that we can be confident that very few batches will have a yield outside the range 62 to 80. Concerning the impurity of future batches we can be sure that none will have less than 0% impurity but we would surely be very reluctant to predict an upper limit. The right hand tail of Fig. 2.2 suggests that the next 50 batches might well contain one with impurity in excess of 12%.

The symmetrical pattern of Fig. 2.1 is a great comfort to both chemist and statistician. A similar pattern is found in many sets of data and we will discuss this shape again in the next chapter. When confronted with a set of data which does *not* conform to this pattern, however, it is always possible to *transform* the data in order to change the shape of their distribution. A suitable transformation may result in a 'better' shape. One very popular transformation is to take logarithms of the individual observations. Taking logs (to the base 10) of the 50 impurity values gives us the 'log impurities' in Table 2.7 and these

Table 2.7   Log impurity of 50 batches

Batch	Impurity	Log impurity	Batch	Impurity	Log impurity
1	1.63	0.21	26	1.78	0.25
2	5.64	0.75	27	2.34	0.37
3	1.03	0.01	28	0.83	−0.08
4	0.56	−0.25	29	1.26	0.10
5	1.66	0.22	30	1.78	0.25
6	1.90	0.28	31	4.92	0.69
7	7.24	0.86	32	1.58	0.20
8	1.62	0.21	33	5.13	0.71
9	2.64	0.42	34	3.02	0.48
10	2.10	0.32	35	0.19	−0.72
11	0.42	−0.38	36	2.00	0.30
12	1.07	0.03	37	1.15	0.06
13	1.52	0.18	38	0.66	−0.18
14	11.31	1.05	39	3.24	0.51
15	2.19	0.34	40	2.24	0.35
16	3.63	0.56	41	2.51	0.40
17	1.07	0.03	42	2.82	0.45
18	3.89	0.59	43	0.37	−0.43
19	2.19	0.34	44	2.63	0.42
20	0.71	−0.15	45	4.17	0.62
21	0.83	−0.08	46	1.38	0.14
22	1.38	0.14	47	3.63	0.56
23	2.45	0.39	48	0.76	−0.12
24	2.34	0.37	49	3.09	0.49
25	8.51	0.93	50	1.29	0.11

Table 2.8   Frequency distribution of log impurity

Log impurity	Number of batches	Log impurity	Number of batches
−0.80 to −0.61	1	0.20 to 0.39	15
−0.60 to −0.41	1	0.40 to 0.59	10
−0.40 to −0.21	2	0.60 to 0.79	4
−0.20 to −0.01	5	0.80 to 0.99	2
0.00 to 0.19	9	1.00 to 1.19	1
		Total	50

transformed values have been used to produce the frequency distribution in Table 2.8.

The frequency distribution of the log impurity values has been represented graphically in Fig. 2.3. Clearly the transformation of the data has altered considerably the shape of the distribution. In Fig. 2.3 we see the symmetrical shape that was noted in Fig. 2.1.

Having admired the symmetry of Fig. 2.3 you might be tempted to ask 'So what? How will this help us to speculate about the impurity of future batches?' We will see in the next chapter that transforming a set of data to get a distribution with this familiar shape may enable us to make use of one of the *theoretical distributions* that statisticians have made available. This can help us to predict what percentage of future batches will have impurity in excess of any particular value.

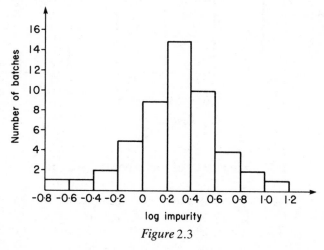

Figure 2.3

The frequency distribution of our third variable (number of overloads) is displayed in Fig. 2.4. This diagram is known as a *bar chart*. You will notice that it differs from the histograms that we have examined previously. In a histogram there are no gaps between blocks or bars whereas in a bar chart we have quite narrow bars with large gaps between them.

The two types of diagram are used to represent different types of variable. Yield and impurity are both *continuous* variables whereas number of overloads is a *discrete* variable. If we have two batches with yields of 70.14 and 70.15 it is possible to imagine a third batch which has yield between these two figures. Since *all* values (in a certain range) are possible we say that yield is a *continuous* variable. On the other hand we may have a batch which suffered one overload during its manufacture and we can have a second batch with two overloads but it would be meaningless to speak of a third batch with 1½ overloads. As there are distinct gaps between the values which 'number of overloads' can have, we say that this is a *discrete* variable.

It is usual to represent the distribution of a continuous variable by a histogram and to represent the distribution of a discrete variable by a bar chart. The gaps between the bars in the bar chart emphasize the gaps between the values that the discrete variable can take.

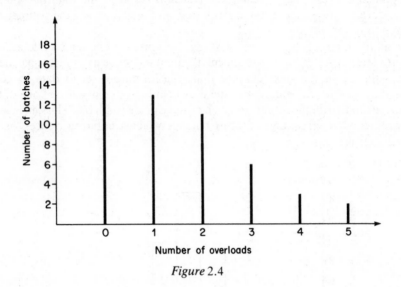

Figure 2.4

2.4   Measures of location and spread

A diagram such as a histogram or a bar chart can give a very clear impression of the distribution of a set of data and the use of diagrams is strongly advocated by statisticians. There is, nonetheless, a need to summarize a distribution in numerical form. This is usually achieved by calculating the mean and the standard deviation of the data.

The *mean* (or arithmetic mean, or average) is easily calculated by adding up the observations and dividing the total by the number of observations. The yields of the first six batches in our sample are:

69.0   71.2   74.2   68.1   72.1   64.3

Adding these six yields we get a total of 418.9 and dividing by 6 we get a mean yield of 69.816 666. Rounding this to two decimal places gives 69.82. (It is usual to round off the mean to one more decimal place than the original data.)

Clearly the calculation of the mean of a set of observations is very straightforward and there is no need to write down a formula for fear of forgetting how the calculation is carried out. We will nonetheless examine a formula as it will introduce a mathematical notation that will be used later.

$$\text{Sample mean} = (\Sigma x)/n \qquad (2.1)$$

$$\text{Population mean} = (\Sigma x)/N \qquad (2.2)$$

In these formulae n is the number of observations in the sample, N is the number of observations in the population and Σx means 'the sum of the observations'.

For those who are not familiar with the use of expressions such as Σx an explanation is offered in Appendix A, and a full list of the symbols used throughout this book is contained in Appendix B.

It is quite possible that you will use equation (2.1) on many occasions but it is doubtful whether you will *ever* use equation (2.2) since you are unlikely to be in the position of having observed every member of a population.

The mean is often referred to as a *measure of location* since it gives us an indication of where a histogram would be located on the horizontal axis. If all the observations in a set of data were increased by adding 10, say, to each then the mean would be increased by 10 and the location of the histogram would be shifted 10 units to the right along the axis.

An alternative measure of location is the *median*. We can find the sample median by listing the observations in ascending order; the median is then the 'middle one'. Listing the yields of the six batches in ascending order gives:

64.3   68.1   69.0   71.2   72.1   74.2

Having listed the six observations we realize that there isn't a 'middle one'. Clearly there will only be a middle observation if the set of data contains an *odd* number of observations. When a set of data contains an *even* number of observations we average the 'middle two' to find the median. Averaging 69.0 and 71.2 we get the median yield to be 70.10.

We have calculated the mean yield to be 69.82 and the median yield to be 70.10. Which is the better measure of location? The answer to this question depends upon what *use* we intend to make of the measure of location and upon the shape of the distribution. Certainly the mean is used more frequently than the median but with a very skewed distribution the median may be preferable. A survey of salaries of industrial chemists would be summarized by quoting a median salary, for example.

A measure of location is often more useful if it is accompanied by a measure of *spread*. In many situations the spread or variability of a set of data is of prime importance. This is particularly true when we are discussing the precision and bias of a test method. The variability or spread of a set of measurements can be used as an indication of the lack of precision of the method or the operator.

The simplest measure of spread is the *range*. The range of a set of data is calculated by subtracting the smallest observation from the largest. For our six batches the range of the yield measurements is 74.2 minus 64.3 which is 9.9.

Though the range is very easy to calculate, other measures of spread are often preferred because the range is very strongly influenced by two particular observations (the largest and smallest). Furthermore the range of a large sample is quite likely to be larger than the range of a small sample taken from the same population. The range, therefore, is not a useful measure of spread if we wish to compare samples of different size. This deficiency is not shared by the two most useful measures of variability:

(a) the standard deviation;
(b) the variance.

We will discuss both at this point since they are very closely related. The standard deviation of a set of data is the square root of the variance. Conversely the variance can be found by squaring the standard deviation. Many electronic calculators have a facility for calculating a standard deviation very easily. If you have such a calculator the easiest way to obtain the variance of a set of data is to first calculate the standard deviation and then square it. On the other hand if you have only a very simple calculator you will find it easier to calculate a variance directly from the formula:

$$\text{Sample variance} = \Sigma(x - \bar{x})^2/(n-1) \qquad (2.3)$$

in which $\bar{x}$ is the sample mean.

The use of this formula will be illustrated by calculating the variance of the yield values for the first six batches. The calculation is set out in Table 2.9.

Table 2.9   Calculation of sample variance by equation (2.3)

	Yield (%) x	Yield − mean $(x - \bar{x})$	(Yield − mean)^2 $(x - \bar{x})^2$
	69.0	−0.82	0.6724
	71.2	1.38	1.9044
	74.2	4.38	19.1844
	68.1	−1.72	2.9584
	72.1	2.28	5.1984
	64.3	−5.52	30.4704
Total	418.9	−0.02	60.3884
Mean	69.82		

$$\text{Sample variance} = \Sigma(x - \bar{x})^2/(n-1)$$
$$= 60.3884/5$$
$$= 12.077\ 68$$
$$\text{Sample standard deviation} = \sqrt{(\text{sample variance})}$$
$$= \sqrt{12.077\ 68}$$
$$= 3.475\ 295\ 6$$

Rounding the results of Table 2.9 to three decimal places gives the sample variance of 12.078 and the sample standard deviation of 3.475. It is a common practice to round off a standard deviation to *two* decimal places more than the original data.

With regard to the calculation in Table 2.9 several points are worthy of mention:

(a) The total of the second column (-0.02) would have been exactly zero if the mean had not been rounded.

(b) It was said earlier that it is a common practice to round off a sample mean to *one* decimal place more than the data. There is no guarantee that a mean rounded in this way will be sufficiently accurate for calculating a standard deviation.

(c) It is also common practice to round off a sample standard deviation to *two* decimal places more than the data. Rounding in this way would give a sample standard deviation of 3.475.

(d) The entries in the third column of Table 2.9 are all positive; therefore the sum of the column [i.e. $\Sigma (x - \bar{x})^2$] must be positive, therefore the sample variance must be positive and the sample standard deviation must be positive.

(e) In equation (2.3) we divide $\Sigma (x - \bar{x})^2$ by $(n-1)$. The question is often asked 'Why do we divide by $(n-1)$ and not by n?' The blunt answer is that dividing by $(n-1)$ is likely to give us a better estimate of the population variance. If we adopt the practice of dividing by n our sample variance will tend to underestimate the population variance.

(f) $\Sigma (x - \bar{x})^2$ is often referred to as the 'sum of squares' whilst $(n-1)$ is referred to as the 'degrees of freedom'. To obtain a variance we divide the sum of squares by its degrees of freedom. Using this terminology we could translate the question in (e) into 'Why does the sum of squares have $(n-1)$ degrees of freedom?' A simple answer can be seen if we refer to the $(x - \bar{x})$ column of Table 2.9. Though this column contains six entries it contains *only five independent entries*. Because the column total must be zero (ignoring rounding errors) we can predict the sixth number if we know any five numbers in the column. If we had n rows in Table 2.9, the centre column would contain $(n-1)$ independent entries and $\Sigma (x - \bar{x})^2$ would have $(n-1)$ degrees of freedom.

Equation (2.3) which we have used to calculate the sample variance is just one of several that we might have used. Equation (2.4) is an alternative which is easier to use especially if a simple calculator is available.

$$\text{Sample variance} = (\Sigma x^2 - n\bar{x}^2)/(n-1) \qquad (2.4)$$

The use of this alternative method will be illustrated by repeating the calculation of the variance of the six yield values which is set out in Table 2.10.

You will note that the value of standard deviation calculated in Table 2.10 (3.394) is not in close agreement with the value calculated earlier (3.475) when using equation (2.3). We noted earlier that the first calculation was in error because we had rounded the mean to two decimal places. This same rounding of the mean has led to a much greater error when using equation (2.4). Had we used a sample mean of 69.816 666 when using equation (2.4) we would have

Table 2.10   Calculation of variance by equation (2.4)

	Yield (%) x	(Yield)^2 x^2
	69.0	4 761
	71.2	5 069.44
	74.2	5 505.64
	68.1	4 637.61
	72.1	5 198.41
	64.3	4 134.49
Total	418.9	29 306.59
Mean	69.82	

$$\text{Sample variance} = (\Sigma x^2 - n\bar{x}^2)/(n-1)$$
$$= [29\ 306.59 - 6(69.82)^2]/5$$
$$= (29\ 306.59 - 29\ 248.994)/5$$
$$= 57.596/5$$
$$= 11.5192$$
$$\text{Sample standard deviation} = \sqrt{(\text{sample variance})}$$
$$= \sqrt{11.5192}$$
$$= 3.394$$

obtained a standard deviation equal to 3.475 294 2 (working with eight decimal places). This agrees favourably with our first result and with the value of 3.475 293 753 calculated on a larger electronic calculator. Perhaps you find these discrepancies alarming.

> Be warned! Carry as many significant figures as your calculator will hold when performing statistical calculations.

The sample standard deviation and the sample variance are closely related and it is very easy to calculate either if we know the other. The important difference between the two measures of spread is the units in which they are measured. The sample standard deviation, like the sample mean, is expressed in the *same* units as the original data. If, for example, the weight of an object has been determined five times then the mean and the standard deviation of the five determinations will be in grams if the five measurements were in grams. The variance of the five determinations would be in grams squared. In some situations we use the sample standard deviation whilst in others we use the sample variance. Whilst the standard deviation can be more easily interpreted than the variance, we are forced to use the latter when we wish to combine measures of variability from several samples, as you will see in a later chapter.

The coefficient of variation is a measure of spread which is dimensionless. To calculate a coefficient of variation we express a standard deviation as a percentage of the mean.

$$\text{Coefficient of variation} = (\text{standard deviation/mean}) \times 100 \quad (2.5)$$

Equation (2.5) has been used to calculate the coefficients of variation for each of the three variables recorded in Table 2.1 and for the log impurities from Table 2.7. The mean, standard deviation and coefficient of variation of each of the four variables are given in Table 2.11.

Table 2.11   Fifty batches of digozo blue pigment

Variable	Yield	Impurity	Log impurity	Number of overloads
Mean	71.23	2.486	0.266	1.5
Standard deviation	3.251	2.1204	0.3488	1.53
Coefficient of variation (%)	4.6	85.3	131	102

Comparing the standard deviations in Table 2.11 we see that the standard deviation in the yield column (3.251) is the greatest of the four. Yield is the most variable of the four variables. Comparing the coefficients of variation we see that the smallest of the four (4.6%) is in the yield column. In *relative* terms, then, yield is the least variable of the four sets of measurements.

The comparisons we have just made in Table 2.11 are rather artificial since it is of no practical importance to be aware that impurity is more variable than yield or vice versa. In this particular situation we are more concerned with comparing the yield and/or impurity of individual batches with limits laid down in a specification. There are many other situations, however, in which the comparison of standard deviations and/or coefficients of variation is of fundamental importance. In the analytical laboratory, for example, we might find that the errors of measurement in a particular method of test tend to be greater when a larger value is being measured. If we obtain repeat determinations at several levels of concentration we would not be surprised to find results like those in Table 2.12.

Table 2.12   Precision of a test method

Level of concentration	1	2	3	4
Determinations	15.1, 15.2, 15.2	29.9, 30.1, 30.0	40.4, 40.3, 40.6	57.3, 57.1, 57.5
Mean determination	15.17	30.00	40.43	57.3
Standard deviation	0.0577	0.1000	0.1528	0.2000
Coefficient of variation (%)	0.38	0.33	0.38	0.35

We can see in Table 2.12 that the standard deviation increases as the level of concentration increases. When, however, the standard deviation is expressed as a percentage of the mean to give the coefficient of variation, we find that this measure of relative precision is substantially constant.

Note that the use of coefficients of variation is restricted to measurements which have a true zero (i.e. ratio scale measurements). When measurements have an arbitrary zero (interval scale), then the coefficient of variation may be meaningless. Suppose, for example, we have three repeat determinations of temperature which could be expressed in degrees centigrade or degrees Fahrenheit, as in Table 2.13. The three Fahrenheit temperatures are exact equivalents of the centigrade temperatures, with no rounding errors being introduced. We see that the coefficients of variation differ and the difference arises because of the 32 degrees *shift in the zero* as we change from one scale to the other.

Table 2.13 Measurements with an arbitrary zero

Temperature scale	Determinations	Mean	SD	C of V (%)
Centigrade	38.0, 40.0, 40.5	39.50	1.323	3.35
Fahrenheit	100.4, 104.0, 104.9	103.1	2.381	2.31

2.5 Cumulative frequency distributions

Earlier we tabulated four frequency distributions to describe the batch to batch variability in yield, impurity, log impurity and number of overloads. From any of these tables we can quickly see how many batches fall into a particular group. In Table 2.2, for example, we see that 10 batches had a yield between 68.0 and 69.9. In Table 2.6 we see that 12 batches had percentage impurity between 2.00 and 2.99%.

For some purposes we need to know how many batches had a yield *less than* a specified value or how many batches had impurity *less than* a certain percentage. In order to furnish such information in a convenient form we tabulate what is known as a *cumulative frequency distribution*. This is produced by adding (or cumulating) the frequencies in a frequency distribution table. To produce a cumulative frequency distribution for the yield of the 50 batches we read Table 2.2 from left to right, adding the frequencies as we progress. Thus we see that:

$$1 \text{ batch had yield less than } 64.0$$

$$(1+2) = 3 \text{ batches had yield less than } 66.0$$

$$(1+2+4) = 7 \text{ batches had yield less than } 68.0$$

Putting this information into a tabular form gives us the cumulative frequency distribution in Table 2.14.

Table 2.14 Cumulative frequency distribution – yield

Yield (%)	62.0	64.0	66.0	68.0	70.0	72.0	74.0	76.0	78.0	80.0
Number of batches having yield less than the above value	0	1	3	7	17	30	39	45	48	50

The numbers in the bottom row of Table 2.14 are known as cumulative frequencies. Note how these cumulative frequencies increase from 0 at the left of the table to 50 at the right. This is very different to the pattern in the bottom row of Table 2.2. If we put Table 2.14 into graphical form we would expect it to give a picture very different from the histogram of Fig. 2.1. This expectation is confirmed by Fig. 2.5.

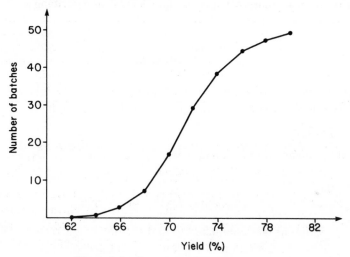

Figure 2.5 Cumulative frequency curve

Fig. 2.5 is known as an 'ogive' or a 'cumulative frequency curve'. The latter name could be misleading as it is more usual to join the points by straight lines rather than by a continuous curve. The distinctive S shape of Fig. 2.5 is associated with the symmetrical histogram that we saw in Fig. 2.1 which represents the frequency distribution of yield. If we draw a cumulative frequency curve for the impurity determinations we would have obtained a rather different shape which is a consequence of the skewness in the impurity distribution.

2.6 Summary

In this chapter we have examined some of the calculations, tabulations and diagrams which can be used to condense a set of data into a more digestible form. Diagrams and tables are invaluable when reporting the results of empirical research. There is some truth in the saying 'A picture says more than a hundred words' but we must not underestimate the summary power of simple calculations.

There are several measures of location and spread but the most commonly used are the *mean* and *standard deviation*. These two will certainly be used repeatedly throughout this book. In later chapters, when we raise our eyes to look beyond the sample towards the population from whence the sample came, we will use the sample mean and standard deviation to estimate the location and spread of the population. These estimates can assist us to draw conclusions about the important features of the population so that we can predict what is likely to occur in the future or select from alternative courses of action.

Before we can devote ourselves exclusively to this main task of 'testing and estimation' we will spend the next chapter discussing ways of describing a population. Unfortunately that is not quite so easy as describing a sample.

Problems

(1) The number of occasions on which production at a complex production plant was halted each day over a period of ten days are as follows:

   4   2   7   3   0   3   1   13   4   3

(a) Find the mean and median number of production halts each day.
(b) Find the variance and standard deviation of the number of production halts each day.
(c) Is the variable measured above discrete or continuous?

(2) An experiment was conducted to find the breaking strength of skeins of cotton. The breaking strengths (in pounds) for a small initial sample were:

   90   99   97   89   108   99   82   96

(a) Find the mean, median, variance, standard deviation and coefficient of variation of this sample.
(b) Is the variable measured above discrete or continuous?
(c) Use the automatic routine on a calculator to confirm your answers for the mean and standard deviation.

(3) Crosswell Chemicals have obtained a long-term order for triphenolite from a paint manufacturer who has specified that the plasticity of the triphenolite must be between 240.0 and 250.0. The first 50 batches of triphenolite produced by Crosswell give the plasticities listed opposite.

   If a batch is above specification it has to be scrapped at a cost of £1000. If it is below specification it can be blended to give a consignment within specification. The cost of blending is £300 for each batch below specification. The process can easily be adjusted to give a different mean level of plasticity and

		Batch number		
1–10	*11–20*	*21–30*	*31–40*	*41–50*
240.36	245.71	248.21	242.16	247.52
245.21	247.32	243.69	250.24	244.71
246.09	244.67	246.42	241.74	246.09
242.39	244.56	242.73	247.71	249.31
249.31	245.61	249.31	245.61	243.28
247.32	241.51	245.32	244.91	247.11
243.04	246.21	244.11	245.72	242.66
245.72	242.01	250.23	246.31	245.91
244.22	246.51	247.01	243.54	246.17
251.51	243.76	240.71	244.62	248.26

any such adjustment would not be expected to change the batch to batch variability of the production process.

The plasticity of the first 50 batches is summarized in the following table and Fig. 2.6:

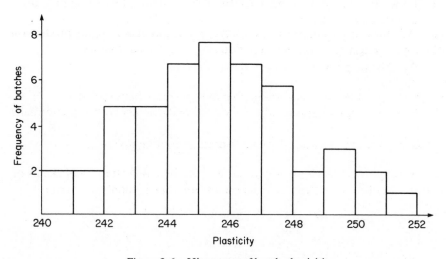

Figure 2.6   Histogram of batch plasticities

Sample size	50
Mean	245.49
Standard deviation	2.561
Median	245.66
Range	11.15
Variance	6.543
Coefficient of variation	1.04%

Grouped frequency table

Plasticity	Number of batches
240.0 and under 241.0	2
241.0 and under 242.0	2
242.0 and under 243.0	5
243.0 and under 244.0	5
244.0 and under 245.0	7
245.0 and under 246.0	8
246.0 and under 247.0	7
247.0 and under 248.0	6
248.0 and under 249.0	2
249.0 and under 250.0	3
250.0 and under 251.0	2
251.0 and under 252.0	1

Using the analysis of data given above:

(a) To what value should the mean be set to minimize losses due to 'out-of-specification' batches?
(b) Do you consider that the sample has been drawn from a population with a symmetrical distribution?
(c) What is the population referred to in part (b)?

—————3—————
Describing the population

3.1 Introduction

In Chapter 2 we examined various ways of describing a sample. The methods we employed were basically very simple but they are, nonetheless, very powerful as a means of condensing a set of data into a more digestible form. You will recall that we made use of:

(a) frequency distributions;
(b) diagrams (such as histograms and bar charts);
(c) measures of location and spread (such as means and standard deviations).

We can adopt a similar approach when we wish to describe a *population*. There are, however, one or two complications which make the description of a population rather more difficult than the description of a sample. Firstly, the population will be much larger and may in some cases be infinitely large. Secondly, we will very rarely be able to measure every member of the population and this will necessitate that we *infer* some of the characteristics of the population from what we have found in the sample.

Statistical inference is so important that it will dominate the later chapters of this book. In this chapter we will confine our attention to several widely used population descriptions, which are known as probability distributions. These probability distributions are necessarily more abstract than the frequency distributions of the previous chapter and you may feel that you are being led astray if your main interest is data analysis. Let it be stated very clearly at the outset, therefore, why it is necessary to study probability distributions as a means of describing populations. There are two main reasons:

(a) We can use a probability distribution to predict unusual events. (When doing so we will be using the probability distribution as a model. Clearly it is important to choose a good model.)
(b) The methods of statistical inference that are described in later chapters are based on probability distributions. It is necessary to have some appreciation of the *assumptions* underlying these methods and this is only possible if you have some knowledge of the well known distributions.

3.2 Probability distributions

When we transfer our attention from the sample to the population we must abandon the frequency distribution and adopt a probability distribution. There are similarities between the two, however, especially if we are describing a discrete variable. It will ease the transition from frequencies to probabilities therefore if we reconsider the frequency distribution in Table 2.5 which is reproduced as Table 3.1.

Table 3.1   Frequency distribution for 50 batches of pigment

Number of overloads	0	1	2	3	4	5	Total
Number of batches	15	13	11	6	3	2	50

Table 3.1 describes the distribution of 'number of overloads' for the 50 batches in the sample. What would this distribution have looked like if we had examined the whole population? Unfortunately we cannot answer this question unless we know exactly what we mean by 'the population'. Clearly the population consists of batches of digozo blue pigment, but which batches and how many batches?

We are free to define the population in any way we like. Perhaps the population should consist of those batches *about which we wish to draw a conclusion*. It would be wise, then, to include *future* batches in our population since the chief chemist's prime concern is to use what has happened in the past to predict what is likely to happen in the future. We will, therefore, define our population to be '*all* batches of digozo blue pigment including past, present and future production'.

Having included future batches we cannot know exactly how many there will be. When speaking of the whole population, then, we will abandon frequencies and use proportions. We could, of course, have used proportions when describing the sample and this is illustrated by Table 3.2.

Table 3.2   Proportional frequency distribution for 50 batches

Number of overloads	0	1	2	3	4	5	Total
Proportion of batches	0.30	0.26	0.22	0.12	0.06	0.04	1.00

In Table 3.2 the numbers in the bottom row are proportions. We can see that the proportion of batches which were subject to two overloads was 0.22, for example. We note that the proportions add up to 1.00, which would be true no matter how many batches we had in the sample. Perhaps, then, something similar to Table 3.2 would be suitable for describing the whole population of batches. We do in fact use a table known as a *probability distribution*. This is based on *probabilities*, which in many respects are very similar to proportions. A formal definition of 'probability' will not be given in this book but the reader can safely regard a probability as a *population proportion*. A set of probabilities can be estimated by calculating a set of sample proportions like those in Table 3.2.

Table 3.3   Probability distribution for the population of batches

Number of overloads	0	1	2	3	4	5	
							Total
Probability							1.00

The probabilities which are missing from Table 3.3 will each be numbers between 0.00 and 1.00, like the proportions in Table 3.2. Furthermore the probabilities will add up to 1.00 which is another similarity between Table 3.2 and 3.3.

How are we to obtain the probabilities which will enable us to complete Table 3.3? One possibility is to estimate them from the sample. This could be done by simply transferring the proportions from Table 3.2 to Table 3.3 and calling them probabilities. This would, of course, give misleading results if the sample were not representative of the population. Another possibility is to make use of one of the well known probability distributions bequeathed to us by mathematical statistics. One of these theoretical distributions which has been found useful in many situations is known as the Poisson distribution.

3.3   The Poisson distribution

Throughout the population the 'number of overloads' will vary from batch to batch. If the 'number of overloads' has a Poisson distribution then we can calculate the required probabilities using:

$$\text{Probability of } r \text{ overloads} = \mu^r e^{-\mu}/r! \qquad (3.1)$$

$r!$ is read as 'r factorial' and is defined as follows:

$$0! = 1, \quad 1! = 1, \quad 2! = 2 \times 1, \quad 3! = 3 \times 2 \times 1, \quad 4! = 4 \times 3 \times 2 \times 1, \text{etc.}$$

The calculation of Poisson probabilities is quite easy on a pocket calculator once we have a value for μ and a value for r. μ is the mean number of overloads for all the batches in the population. Clearly we do not know the value of μ but we can estimate it from the sample. As the mean number of overloads for the 50 batches in the sample is 1.50 we will let $\mu = 1.5$. Letting $r = 0$ we can now calculate:

$$\text{Probability of 0 overloads} = (1.5)^0 e^{-1.5}/0!$$

$$= 0.2231$$

(*Note* that $(1.5)^0 = 1$.)

This result is telling us that, if we select a batch at random from the population, there is a probability of 0.2231 that 0 overloads will occur during its production. Most people prefer percentages to probabilities so we could say, alternatively, that 22.3% of batches in the population would be subjected to 0 overloads during manufacture.

Letting $r = 1$, then letting $r = 2$, etc. in equation (3.1), we can calculate:

$$\text{Probability of 1 overload} = (1.5)^1 e^{-1.5}/1! = 0.3347$$

$$\text{Probability of 2 overloads} = (1.5)^2 e^{-1.5}/2! = 0.2510$$

$$\text{Probability of 3 overloads} = (1.5)^3 e^{-1.5}/3! = 0.1255$$

$$\text{Probability of 4 overloads} = (1.5)^4 e^{-1.5}/4! = 0.0471$$

$$\text{Probability of 5 overloads} = (1.5)^5 e^{-1.5}/5! = 0.0141$$

These probabilities could now be inserted in Table 3.3. Unfortunately the completed table would be a little ambiguous since the probabilities would not add up to 1.0. The total of the calculated probabilities is only 0.9955.

The missing 0.0045 is explained by the fact that the Poisson distribution does not stop at $r = 5$. We can insert higher values of r into equation (3.1) to obtain more probabilities. Doing so gives:

$$\text{Probability of 6 overloads} = 0.0035$$

$$\text{Probability of 7 overloads} = 0.0008$$

$$\text{Probability of 8 overloads} = 0.0001$$

The calculations have been discontinued at $r = 8$ because the probabilities have become very small though in theory the Poisson distribution continues forever. Tabulating the probabilities gives us the probability distribution in Table 3.4.

Table 3.4   Poisson distribution with $\mu = 1.5$

Number of overloads	0	1	2	3	4	5	6	7 or more
Probability	0.223	0.335	0.251	0.126	0.047	0.014	0.004	0.001

How do the probabilities in Table 3.4 compare with the proportions in Table 3.2? Perhaps the best way to compare the two distributions is by means of a graph such as Fig. 3.1.

It is quite clear that the two distributions portrayed in Fig. 3.1 are not *identical*. Perhaps it is unreasonable to expect them to be exactly the same. Whenever we take a sample from a population we must not expect that the scatter of values in the sample will have a distribution that is a perfect replica of the population distribution. Just how much variability we can expect from sample to sample will be discussed in a later chapter; at this point we will simply note that a sample is most unlikely to resemble in every respect the population from which it was taken.

So the two distributions in Fig. 3.1 are different, but not *very* different. If someone put forward the suggestion that 'the population of batches had a Poisson distribution with $\mu = 1.5$' would you reject this suggestion on the evidence of the 50 batches in the sample? It is very doubtful!

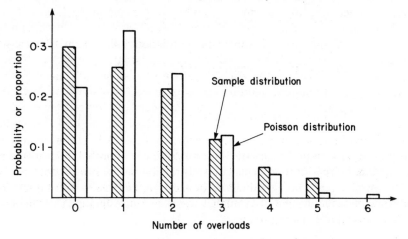

Figure 3.1   Multiple bar chart to compare two distributions

Suppose, then, we accept that the number of overloads has a Poisson distribution with $\mu = 1.5$, what conclusions can we draw concerning future batches? We can use the probabilities in Table 3.4 to predict what percentage of future batches will be subjected to any particular number of overloads. We could predict, for example, that 1.4% of batches (i.e. 0.014×100) will have five overloads and that 0.4% of batches will have six overloads. By adding probabilities from Table 3.4 we can predict that 1.9% of batches [i.e $(0.014 + 0.004 + 0.001) \times 100$] will have been subjected to at least five overloads during manufacture.

The *importance* of such predictions will depend on whether or not the occurrence of overloads is causing a reduction in yield and/or an increase in impurity. The chief chemist has suggested that interruptions in agitation have caused excess impurity, but this has yet to be established.

The *accuracy* of the predictions will depend upon the applicability of the Poisson distribution to this situation. The Poisson distribution with $\mu = 1.5$ certainly matches the sample distribution quite well, but there are many other distributions which could have been tried, so why was the Poisson selected? It is well known that certain probability distributions are useful in certain situations and the Poisson distribution has proved useful for describing the occurrence of equipment failures. Further support for its use would be found in the mathematical foundation of the Poisson distribution. In the derivation of the Poisson probability formula [equation (3.1)] it is assumed that events occur at random in a continuance such as time, or a line, or a space, etc. It is further assumed that the events occur independently of each other and at a fixed rate per unit time or per unit length, etc. In any succession of intervals the number of events observed will vary in a random manner from interval to interval with the probability of *r* events occurring in any particular interval being given by equation (3.1). The Poisson distribution has been found by the textile technologist to account for the scatter of faults along a continuous length of yarn and has been found useful by geographers when attempting to explain the scatter of certain natural phenomena in a continuous area of land. The decision to use the Poisson distribution, rather than an alternative, could have serious consequences, of course. Only if we select a *suitable* model can we hope to make accurate predictions of unlikely events such as 'eight overloads occurring during the manufacture of a batch'. You would need to know much more about several *other* distributions, before you were able to select the 'best' distribution for this task.

It is not our intention to discuss the finer points of probability distributions in this book. The reader who wishes to study further should consult Johnson and Leone, Vol. 1 (1964), Freund (1962) or Moroney (1966). We must now leave the Poisson distribution and examine what is undoubtedly the most important of all probability distributions.

3.4   The normal distribution

We have referred to 'number of overloads' as a *discrete* variable. Clearly the number of overloads occurring during the manufacture of any batch can only be a whole number (i.e. 0, 1, 2, 3 etc.) and when using equation (3.1) to calculate Poisson probabilities we substituted whole number values for *r*. The other two variables that were measured on each of the 50 batches, yield and impurity, are *not* discrete variables. It would be most inappropriate to attempt to use the Poisson distribution to describe these *continuous* variables.

You will recall that we used histograms to describe the sample distributions for yield and impurity in contrast with the bar chart used to describe the distribution of 'number of overloads'. We noted at the time that one of the histograms (yield) was roughly symmetrical whilst the other was rather skewed. The symmetrical histogram is reproduced in Fig. 3.2.

In Fig. 3.2 a smooth curve has been superimposed on the histogram. Clearly the curve and the histogram are similar but not identical. The histogram represents the distribution of the yield values for the 50 batches in the *sample*. It is suggested that the curve might represent the distribution of the yield values of all the batches in the *population*.

The curve in Fig. 3.2 represents a probability distribution which is known as the *normal distribution*. To be more precise it is a normal distribution which has mean equal to 71.23 and standard deviation equal to 3.251, these values having been chosen to match the mean and standard deviation of the 50 yield values in the sample. You would probably agree that it is reasonable to use the sample

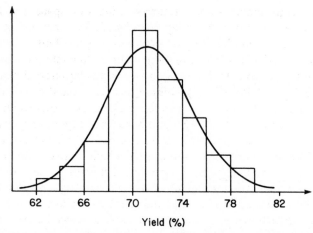

Yield (%)

Figure 3.2   Comparing a histogram with a normal distribution curve

mean and standard deviation as estimates of the population mean and standard deviation, but you may wonder why we should introduce this particular curve. The main reasons why we turn to the normal distribution to describe this population are:

(a) The normal distribution has been found useful in many other situations.
(b) Tables are readily available which allow us to make use of the normal distribution without performing difficult calculations.
(c) The normal distribution can be shown to apply to any variable which is subjected to a large number of 'errors' which are additive in their effects. (If any one source of 'error' becomes dominant we may well get a skewed distribution, as we found with the impurity determinations.)
(d) It is reasonable to expect a *single peak* in the distribution curve since the operating conditions of the plant were chosen to give maximum yield. Any 'error' in the setting of these conditions would result in a loss of yield and this loss might be expected to be roughly equal for a departure in either direction. We would, therefore, expect a roughly *symmetrical* distribution curve.

The normal distribution has been studied by mathematicians for nearly 200 years and the equation of the normal curve also arises in the study of other physical phenomena such as heat flow. It has been found to be very useful for describing the distribution of errors of measurement and many 'natural' measurements are found to have distributions which can be closely approximated by the normal curve. The heights of adult males and females, for example, are said to have normal distributions as illustrated in Fig. 3.3. The two distributions have different means (69 inches and 63 inches) but the same standard deviation (3 inches).

Scores on intelligence tests are also said to have normal distributions. Tests are standardized so that the mean score for all adults is 100. The mean score for

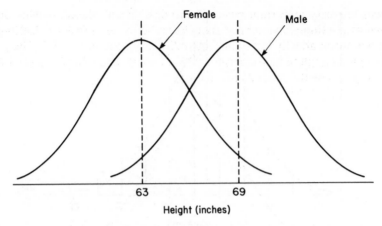

Figure 3.3   Two normal distributions with equal standard deviations but different means

males and the mean score for females are found to be equal but the two distributions are not identical as the standard deviation for males (17) is greater than that for females (13). The two distributions are illustrated in Fig. 3.4.

You will notice that in Fig. 3.4 the female scores are not so widely spread as the male scores. Note also that the narrower of the two distributions is also taller. This must be so because the *areas* under the two curves are equal. In fact the area under each curve is equal to 1.0.

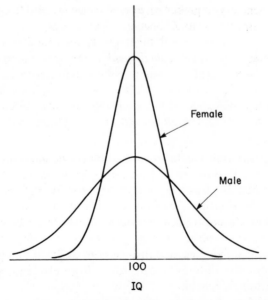

Figure 3.4   Two normal distributions with equal means but different standard deviations

Areas are very important when we are dealing with the probability distributions of continuous variables. By calculating the area under a distribution curve we automatically obtain a probability as we see in Fig. 3.5. The probability that an adult female in Britain will have height between 65 inches and 68 inches is given by the shaded area in Fig. 3.5.

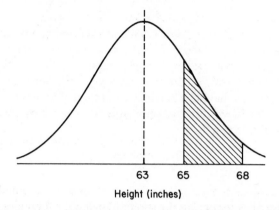

Height (inches)

Figure 3.5 Females with heights between 65 and 68 inches

To find the shaded area in Fig. 3.5 we could resort to mathematical integration but this is completely unnecessary if we make use of Statistical Table A. To obtain the probability of a height lying between 65 and 68 inches we proceed in two stages as follows. In stage 1 we concentrate on those females who are taller than 68 inches. The probability of a female having a height greater than 68 inches is represented by the shaded area in Fig. 3.6(a).

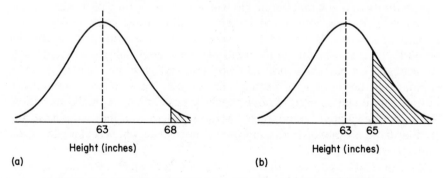

(a) Height (inches)   (b) Height (inches)

Figure 3.6 Females with heights greater than (a) 68 and (b) 65 inches

To find this shaded area we must convert the height (x) into a *standardized* height (z) using:

$$z = \frac{x - \mu}{\sigma}$$

where μ is the population mean and σ is the population standard deviation.

$$z = \frac{68 - 63}{3}$$

$$= 1.67$$

Turning to Table A we find that a z value of 1.67 gives us a probability of 0.0475. Thus, if we select an adult female at random the probability that her height will exceed 68 inches is 0.0475. Alternatively we might say that 4.75% of adult females in Britain are taller than 68 inches.

In stage 2 we concentrate on those females who are taller than 65 inches. The probability of a female having a height greater than 65 inches is represented by the shaded area in Fig. 3.6(b). To find this probability we must once more calculate a standardized height:

$$z = \frac{x - \mu}{\sigma} = \frac{65 - 63}{3} = 0.67$$

Referring to Table A we obtain the probability 0.2515. Multiplying this by 100 leads us to the conclusion that 25.15% of adult females are taller than 65 inches.

Returning to our original problem, it is the shaded area in Fig. 3.5 that we wish to find. This area represents the probability of an adult female having a height between 65 and 68 inches. We can now obtain this result by finding the difference between the two probabilities represented in Fig. 3.6(a) and (b):

$$0.2515 - 0.0475 = 0.2040$$

If, therefore, an adult female is selected at random from the population there is a probability of 0.204 that her height will be between 65 and 68 inches.

Using Table A we could obtain other interesting probabilities concerning the heights of females. We could also make use of Table A to draw conclusions about the heights of males or the intelligence scores of males or females. Table A represents a particular normal distribution which has a mean of 0, and a standard deviation of 1. It is often referred to as the 'standard normal distribution' and it can be used to draw conclusions about *any* normal distribution. When we calculate a standardized value (z) we are in fact jumping *to* the standardized normal distribution *from* the normal distribution in which we are interested.

By using Table A we can draw general conclusions which are applicable to *all* variables which have a normal distribution. For example, we see that a z value of 1.96 in Table A corresponds to a probability of 0.025, which is 2½% if expressed as a percentage. This information is displayed in Fig. 3.7(a) which also shows that 2½% of the area lies to the left of -1.96. Thus we can conclude that 95% of z values lie between -1.96 and $+1.96$. Transferring our attention to Fig. 3.7(b), which represents a normal distribution with mean equal to μ and standard deviation equal to σ, we can say that 95% of values will lie between

$(\mu - 1.96\sigma)$ and $(\mu + 1.96\sigma)$. Without using mathematical symbols we can conclude that:

> For any variable which has a normal distribution, 95% of values will be within 1.96 standard deviations of the mean.

Now that we appreciate the wide applicability of Table A and of the standard normal distribution let us return to our batches of digozo blue pigment. After

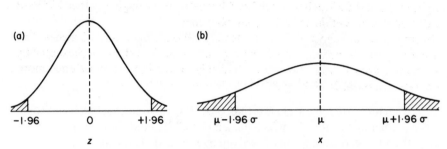

Figure 3.7 Central 95% of a normal distribution

inspecting Fig. 3.2 you may be prepared to accept that the whole population of batches could have a normal distribution. If you are also prepared to accept that the mean of this distribution is approximately 71.2 and the standard deviation approximately 3.25 we can predict the yield we are likely to get in future batches. (These estimates of the population mean and standard deviation are based on the sample mean and standard deviation. Estimation will be discussed fully in later chapters.) Using $\mu = 71.2$ and $\sigma = 3.25$ we calculate:

$$\mu + 1.96\sigma = 71.2 + 1.96\,(3.25) = 77.4$$
$$\mu - 1.96\sigma = 71.2 - 1.96\,(3.25) = 65.0$$

We can therefore expect that 95% of future batches will have yield between 65.0 and 77.4.

If we wish to quote an interval that will embrace 99% of batches we can make use of the following result:

> For any variable which has a normal distribution 99% of values will lie within 2.58 standard deviations of the mean.

$$\mu + 2.58\sigma = 71.2 + 2.58\,(3.25) = 79.6$$
$$\mu - 2.58\sigma = 71.2 - 2.58\,(3.25) = 62.8$$

Thus we can expect that 99% of future batches will have yield between 62.8 and 80.2.

You may recall that the chief chemist was particularly concerned about those batches with low yield and he wished to estimate the percentage of future

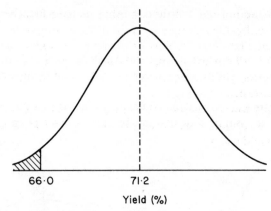

Figure 3.8   Batches of pigment with yield less than 66.0

batches which will have yield less than 66.0. If we assume that batch yields came from a normal distribution which has a mean of 71.2 and a standard deviation of 3.25 then the estimate we need is given by the shaded area in Fig. 3.8.

Following the usual procedure we will first calculate a standardized yield then refer to Table A.

$$z = \frac{x - \mu}{\sigma} = (66.0 - 71.2)/3.25 = -1.60$$

The standardized value is *negative* because we have standardized a yield value (66.0) which is *below* the mean. Table A does not contain any negative z values because it is a tabulation of only the right hand side of the standard normal distribution. Because of the symmetry of the distribution there is no need to tabulate both sides. When we calculate a z value which is negative we simply ignore the minus sign whilst using Table A. Following this practice we get a probability of 0.0548 from Table A. Thus we conclude that 5.48% of future batches will have yield less than 66.0 if the present situation continues.

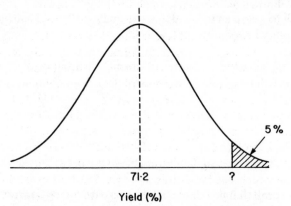

Figure 3.9   Yield exceeded by the top 5% of batches

In our use of the normal distribution table we have first converted a value into a standardized value and then referred to Table A in order to convert this standardized value into a probability or a percentage. Occasionally we use the normal distribution table in the opposite direction. Suppose, for example, we wish to estimate the yield value that will be exceeded by only 5% of batches. The problem is illustrated in Fig. 3.9.

To find the unknown value of yield (x) we enter Table A with a probability (5% = 0.0500) to obtain a standardized value ($z = 1.64$). This must now be converted to a yield using:

$$z = \frac{x - \mu}{\sigma}$$

$$1.64 = \frac{x - 71.2}{3.25}$$

$$x = 71.2 + 1.64 \, (3.25)$$

$$x = 76.53$$

Thus we can expect that only 5% of future batches will have yield greater than 76.53.

3.5   Normal probability graph paper

Our decision to use the normal distribution as a model was based entirely on Fig. 3.2. The histogram, which describes the 50 batches in the sample, closely resembles the normal curve so we conclude that the normal curve might well describe the population of batches. The normal distribution in Fig. 3.2 has the same mean and standard deviation as the histogram.

Superimposing a normal curve onto a histogram is more difficult than you might imagine. In fact the task is so difficult that in practice an alternative approach is used. This approach involves plotting the sample data onto special graph paper known as 'normal probability paper'. In order to use this special paper we must return to the percentage cumulative frequency distribution that was tabulated in Chapter 2 and is reproduced as Table 3.5.

In Chapter 2 we used the pairs of numbers in the above table to plot a cumulative frequency curve. At the time we commented on the distinctive 'ogive' shape of the graph. If we now plot the same points on 'normal probability paper' we get a very different shape, as seen in Fig. 3.10. (Note that the percentages in Table 3.5 have each been reduced by 1% before plotting. For further details of normal probability plots; see Chatfield, 1978.)

Plotting the cumulative frequency distribution on 'normal probability paper' has straightened out the ogive shape so that the points lie almost on a straight line. The line which has been drawn on Fig. 3.10 corresponds to the normal distribution curve in Fig. 3.2; clearly it fits the data very well.

You may recall that the distribution of the impurity determinations of the 50

Table 3.5   Percentage cumulative frequencies – yield

Yield (x)	Number of batches with yield less than x	Percentage of batches with yield less than x
64.0	1	2
66.0	3	6
68.0	7	14
70.0	17	34
72.0	30	60
74.0	39	78
76.0	45	90
78.0	48	96
80.0	50	100

batches was very skewed. If we were to plot the cumulative frequency distribution of impurity on normal probability paper we would not expect to get a straight line as we did for the batch yields. The point is illustrated by Fig. 3.11.

Plotting a cumulative frequency distribution on normal probability paper can give a useful indication of whether or not the data came from a population which has a normal distribution. If we feel that the points would fit closely to a straight line we can draw in a suitable line and then proceed to estimate the mean and standard deviation of the population. The procedure is illustrated in Fig. 3.12.

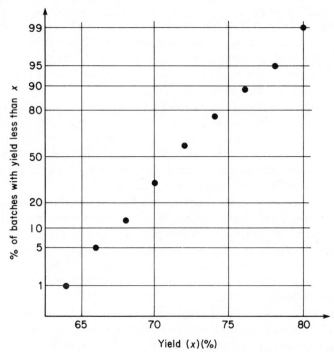

Figure 3.10   Percentage cumulative frequencies plotted on normal probability paper

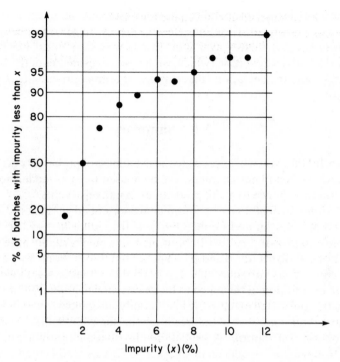

Figure 3.11   Cumulative distribution of impurity on normal probability paper

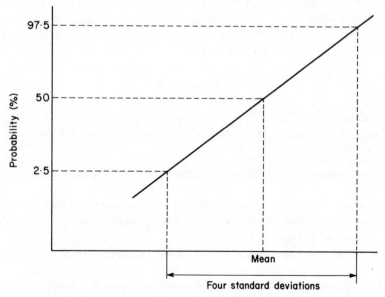

Figure 3.12   Estimation of mean and standard deviation from normal probability paper

The use of normal probability paper to estimate the mean and standard deviation of a normal distribution offered a great saving in effort before pocket calculators became readily available. It is now so easy to calculate a sample mean and standard deviation (which can be used as estimates of the population mean and standard deviation) that this use of normal probability paper is not important.

3.6   Summary

It is doubtful that you will appreciate the usefulness of probability distributions until you have read later chapters. We have used them as models to describe populations and hence to make predictions of unlikely events.

Many other probability distributions are described in standard statistical texts such as Johnson and Leone (1964). If the author had wished to devote more space to probability distributions then he would certainly have discussed the log-normal distribution and the exponential distribution. Both are applicable to continuous variables. The log-normal distribution might well give an adequate description of the batch to batch variation in impurity. The exponential distribution is used extensively in reliability assessment since it is applicable to the time intervals between failures when the number of failures in an interval has a Poisson distribution. A probability distribution known as the binomial distribution will be introduced in Chapter 6. This also has certain connections with the Poisson distribution.

An extremely important use of probability distributions is in the production of the statistical tables which we will use in significance testing and estimation. Though you will never produce a statistical table yourself it is most important that you have an awareness of the assumptions underlying these tables and the statistical techniques which they support.

Problems

(1) A manufacturer knows from experience that the resistance of resistors he produces is normally distributed with mean $\mu = 80$ ohms and standard deviation $\sigma = 2$ ohms.

(a) What percentage of resistors will have resistance between 78 ohms and 82 ohms?
(b) What percentage of resistors will have resistance of more than 83 ohms?
(c) What is the probability that a resistor selected at random will have resistance of less than 79 ohms?
(d) Between what symmetrical limits will the resistance of 99% of resistors lie?

(2) The skein strength of a certain type of cotton yarn varies from piece to piece to the pattern of the normal distribution, with a mean of 96 lb and a standard deviation of 8 lb.

(a) What percentage of skeins will have a strength in excess of 100 lb?
(b) What is the probability that a randomly chosen skein has a strength of between 90 lb and 100 lb?
(c) What strength will be exceeded by 20% of the skeins?
(d) (More difficult.) What is the probability that *two* randomly chosen skeins will *both* have a strength of less than 100 lb? (You will need to make use of the 'multiplication rule for probabilities': probability of A *and* B = (probability of A) × (probability of B) where A and B are independent events.)

(3) A manufacturer of paper supplies a particular type of paper in 1000 metre rolls. It has been agreed with a regular customer that a particular defect known as 'spotting' should not occur more than once per 10 metres on average. The manufacturer feels confident that he can meet this specification since his records show that spotting occurs at a rate of once per 15 metres on average when his production process is working normally.

(a) A 15 metre length of paper is taken from a roll produced under normal conditions. What is the probability that this length of paper will contain no occurrences of spotting?
(b) A 30 metre length of paper is taken from a roll produced under normal conditions.

   (i) What is the probability that this length of paper will contain no occurrences of spotting?
   (ii) What is the probability that this length of paper will contain more than four occurrences of spotting?

(c) (More difficult.) The regular customer decides to check the quality of each roll by inspecting *two* 30 metre lengths selected at random from the 1000 metres on the roll. If he finds fewer than five faults in *each* of the two pieces he intends to accept the roll but if he finds five or more faults in *either or both* he will subject the roll to full inspection.

   (i) What is the probability that a roll produced under normal conditions will be subjected to full inspection? (You will need the multiplication rule again.)
   (ii) What is the probability that a roll with a fault rate of one occurrence of spotting per 5 metres of paper will avoid full inspection?

(4) The Weights and Measures Act of 1963 specifies that prepacked goods which have a weight marked on the package should conform to certain restrictions. The Act specifies that no prepack should have a weight which is less than the nominal weight (i.e. the weight marked on the package). Smith and Co. manufacture cornflour which is sold in cartons having a nominal weight of 500 grams. The packaging is carried out by sophisticated automatic machines which, despite their great cost, are *variable* in performance. The mean weight

of cornflour per packet can be adjusted of course but the variability from packet to packet can not be eliminated and this variability gives rise to a standard deviation of 10 grams regardless of the mean weight at which the machine is set.

(a) If the weights dispensed into individual packages have a normal distri-
    bution with a standard deviation of 10 grams what setting would you
    recommend for the mean weight dispensed in order to comply with the
    1963 Act?

The *Guide to Good Practice in Weights and Measures in the Factory*, published by the Confederation of British Industry in 1969, is regarded by manufacturers and by local authority inspectors as an amendment to the 1963 Act. This document relaxes the restrictions in the Act by means of two concessions:

Concession 1 permits a 1 in 1000 chance of a package being grossly deficient (i.e. having a net weight which is less than 96% of the nominal weight).
Concession 2 permits a 1 in 40 chance of a package being deficient (i.e. having a net weight less than the nominal weight).

(b) What is the minimum setting for the process mean weight which will satisfy
    Concession 1?
(c) What is the minimum setting for the process mean weight which will satisfy
    Concession 2?
(d) What setting would you recommend for the mean weight in order to satisfy
    both concessions?

On 1 January 1980 new regulations came into force which replace the 'minimum system' of the 1963 Act with an 'average system' which conforms to EEC recommendations. This new system is perhaps best described by quoting the three 'packers' rules' on which it is based:

Rule 1: The contents must not be less on average than the nominal weight.
Rule 2: The content of the great majority of prepacks (97.5%) must lie above the tolerance limit derived from the directive. (Tolerance limit is defined as 'nominal weight minus tolerance' and the tolerance specified for a package with weight around 500 grams is 7.5 grams.)
Rule 3: The contents of every prepack must lie above the absolute tolerance limit. (Absolute tolerance limit is defined as 'nominal weight minus twice the tolerance'.)

(e) If Smith and Co. continue to pack with the process mean set at the figure
    recommended in (d) will they comply with Rules 1, 2 and 3?
(f) Assuming that the phrase 'every prepack' in Rule 3 should read '99.99% of
    prepacks' would the process mean setting recommended in (d) lead to
    violation of Rules 1, 2 and 3?
(g) Assuming the modification to Rule 3 suggested in (f), what is the lowest
    value of process mean that will allow all three rules to be satisfied?

4

Testing and estimation: one sample

4.1 Introduction

Perhaps the previous chapter was a digression from the main purpose of this book. No matter how interesting you found the content of Chapter 3 you may have felt that it was a diversion from the main theme of *using data from a sample to draw conclusions about a population.* Even after studying Chapter 3 it is probably not entirely clear how an understanding of the normal distribution can help us to answer some of the questions posed in Chapter 2, questions to which we must now return.

You will recall that the manufacture of digozo blue pigment had been interrupted whilst modifications were made to the plant. The chief chemist is anxious to know whether the mean yield of the plant has increased as a result of these modifications. He is also concerned to check that any increase in yield which might have taken place has *not* been accompanied by an increase in impurity.

In this chapter, then, we are going to talk about the performance of the plant *on average.* We are not going to concern ourselves unduly with individual batches of pigment. We will in fact focus our attention on three questions:

(a) Has the mean yield of the plant increased since the modifications were carried out?
(b) If the mean yield *has* increased, what is its new value?
(c) When we want to detect an increase in yield, how many batches do we need to inspect?

4.2 Has the yield increased?

Before the modifications were carried out the mean yield was 70.30. This figure was obtained by averaging the yield from many hundreds of batches produced during a two-year period when the production process was operating 'satisfactorily'. Not *all* the batches produced during this period were included when calculating the mean. A small number of batches were excluded because they were thought to be 'special cases'. In each of the excluded batches the yield was very low and/or the impurity was very high; we will discuss these 'rogue batches' more fully when we talk about outliers in a later chapter.

So the mean yield was 70.30 *before* the modifications were carried out and the mean yield is 71.23 for the batches produced *after* the modifications. Can we conclude then that the incorporation of the modifications has been effective in increasing the yield? You may be tempted to answer simply, 'yes'. Before doing so, however, it would be wise to take account of the fact that the question refers to a *population* of batches whilst the mean yield of 71.23 was obtained from only a *sample*.

When the chief chemist asks the question 'Has the mean yield increased?' he is surely thinking *beyond* the 50 batches produced since the modification, to the batches which might be produced in the future. In attempting to answer this question, then, we must regard the 50 batches as being a sample from a population which consists of *all* batches made after the modification, including future batches. The question can then be reworded to read 'Is the population mean greater than 70.30?'

The only evidence available to help us answer this question is the yield values of the 50 batches in the sample. You might imagine therefore that the mean yield of future batches is simply a matter of opinion and that one man's opinion is as good as any other. If this were so the evaluation of the effectiveness of a modification would be very *subjective* indeed. Fortunately, however, statisticians have devised a technique known as significance testing (or hypothesis testing) which is purely *objective* and can be used to answer many of the questions we have asked. Before we carry out a significance test let us decide what factors will need to be taken into account. We will do this by examining two hypothetical samples of batches.

Suppose that the chief chemist was anxious to make an early decision concerning the effectiveness of the modifications and he decided to draw a conclusion about the mean yield of the modified plant after producing *only six* batches. His conclusion would depend upon what he found when the yield of each batch was measured.

Suppose the six yield values were:

Sample A: 72.1   69.0   74.3   71.2   70.1   70.7

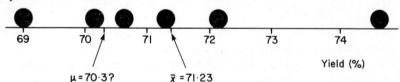

Can we conclude with confidence that the *population* mean yield is greater than 70.3? It is true that the *sample* mean yield (71.23) is greater than this figure but one doubts if the chief chemist would risk his reputation by proclaiming categorically that the mean yield had increased since the modifications were incorporated. Perhaps he would feel that it would be wise to examine further batches before reaching a decision.

Suppose, on the other hand, the first six batches had given the following yield measurements:

Sample B: 71.3   70.7   71.6   72.0   71.1   70.7

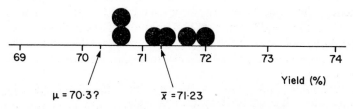

With these results the chief chemist might feel quite strongly that the mean yield *had* increased. He could not be certain, of course, but he might have sufficient confidence to announce that the modifications had effectively increased the yield.

In both of these hypothetical sets of data the mean is 71.23. Thus in both cases the sample mean differs from the old mean (70.30) by 0.93, which is a difference of 1.3%. Yet, with sample B we conclude that the mean yield has increased whereas with sample A we prefer to reserve judgement. What is the essential difference between these two samples which leads to the different conclusions? As we can see by comparing the two blob charts, the batch yields are *much more widely scattered* in sample A than in sample B. This point is confirmed if we compare the standard deviations of samples A and B given in Table 4.1.

Table 4.1   Variation in yield in three samples

Sample	Number of batches n	Sample mean $\bar{x}$	Sample SD s
Hypothetical A	6	71.23	1.828
Hypothetical B	6	71.23	0.5125
Taken by chief chemist	50	71.23	3.251

When trying to decide whether or not the mean yield has increased it is not sufficient to calculate the difference between the sample mean and the old mean; *we must compare this difference with some measure of the batch to batch variation* such as the sample standard deviation.

We will now carry out a significance test using the yield data from the 50 batches inspected by the chief chemist. We will make use of the sample mean ($\bar{x} = 71.23$), the old population mean ($\mu = 70.30$) and the sample standard deviation ($s = 3.251$). As this standard deviation is even larger than those from samples A and B (see Table 4.1) you might expect that we will be unable to conclude that the mean has increased. We will, however, be taking into account the sample size ($n = 50$) and this will have a considerable influence on our decision.

One important step in carrying out a significance test is to calculate what is known as a *test statistic*. For this particular test, which is often referred to as a 'one-sample *t*-test', we will use the following formula.

> *One-sample t-test*
>
> Test statistic $= \dfrac{|\bar{x} - \mu|}{s/\sqrt{n}}$

The 'one-sample *t*-test' is carried out using a six-step procedure which we will follow in many other types of significance tests.

Step 1: Null hypothesis – The mean yield of *all* batches after the modification is equal to 70.30 ($\mu = 70.30$).

Step 2: Alternative hypothesis – The mean yield of *all* batches after the modification is greater than 70.30 ($\mu > 70.30$).

Step 3: Test statistic $= \dfrac{|\bar{x} - \mu|}{s/\sqrt{n}}$

$$= \frac{|71.23 - 70.30|}{3.251/\sqrt{50}}$$

$$= 2.022$$

Step 4: Critical values – From the *t*-table (Statistical Table B) using 49 degrees of freedom for a one-sided test:

1.68 at the 5% significance level
2.41 at the 1% significance level.

Step 5: Decision – As the test statistic lies between the two critical values we reject the null hypothesis at the 5% significance level.

Step 6: Conclusion – We conclude that the mean yield is greater than 70.30.

As a result of following this six-step procedure the chief chemist draws the conclusion that the mean yield *has* increased since the modifications were incorporated. He hopes that he has made the right decision in rejecting the null hypothesis but he realizes that he is running a risk of making a wrong decision every time he carries out a significance test. Whatever procedure he follows the risk will be there. It is unfortunately true that whenever we draw a conclusion about a population using data from a sample we run a risk of making an error. One advantage of the six-step procedure is that the risk can be quantified even though it cannot be eliminated. In fact the probability of wrongly rejecting the null hypothesis is referred to as the 'significance level'.

It might be interesting to return to the two sets of hypothetical data and apply the significance testing procedure. Would we reach the same conclusions that were suggested earlier, when we were simply using common sense? Carrying out a one-sample *t*-test on the six yield values in *sample A* gives:

Null hypothesis $\qquad\qquad \mu = 70.30$

Alternative hypothesis $\qquad \mu > 70.30$

$$\text{Test statistic} = \frac{|\bar{x} - \mu|}{s/\sqrt{n}}$$

$$= \frac{|71.23 - 70.30|}{1.828/\sqrt{6}}$$

$$= 1.25$$

Critical values – From the *t*-table using 5 degrees of freedom for a one-sided test:

2.02 at the 5% significance level
3.36 at the 1% significance level.

Decision – As the test statistic is less than both critical values we cannot reject the null hypothesis.

Conclusion – We are unable to conclude that the mean yield is greater than 70.30.

Thus the chief chemist would not be able to conclude that the modifications had been effective in increasing the yield of the plant. This conclusion is in agreement with that suggested earlier, by a visual inspection of the data.

Turning to the data from *sample B*, which previously led us to a very different conclusion, we carry out a one-sample *t*-test as follows:

Null hypothesis $\qquad\qquad \mu = 70.30$

Alternative hypothesis $\qquad \mu > 70.30$

$$\text{Test statistic} = \frac{|\bar{x} - \mu|}{s/\sqrt{n}}$$

$$= \frac{|71.23 - 70.30|}{0.5125/\sqrt{6}}$$

$$= 4.44$$

Critical values – From the *t*-table using 5 degrees of freedom for a one-sided test:

2.02 at the 5% significance level
3.36 at the 1% significance level.

Decision – As the test statistic is greater than both critical values we reject the null hypothesis at the 1% significance level.

Conclusion – We conclude that the mean yield is greater than 70.30.

This conclusion is again in agreement with that suggested by a visual examination of the data. In many cases the significance testing procedure will lead us to the same conclusion that would have been reached by any reasonable person. It is in the borderline case that the significance test is most useful.

4.3   The significance testing procedure

Carrying out a one-sample *t*-test is quite straightforward if we stick to the recommended procedure. Understanding the theory underlying a one-sample *t*-test is much more demanding, however, and few chemists who carry out significance tests have a *thorough* understanding of the relevant statistical theory. Fortunately, a really deep understanding is not necessary. We will discuss significance testing theory to some extent in the later chapters and a number of comments are offered at this point in the hope that they will make the procedure seem at least reasonable.

(1) The *null hypothesis* is a simple statement about the population. The purpose of the significance test is to decide whether the null hypothesis is true or false. The word 'null' became established in significance testing because the null hypothesis often specifies *no* change, or *no* difference. However, in some significance tests the word 'null' appears to be most inappropriate, and a growing number of statisticians prefer the phrase 'initial hypothesis'. We will continue to use 'null hypothesis' throughout this book.

(2) The *alternative hypothesis* also refers to the population. The two hypotheses, between them, cover all the possibilities we wish to consider. Note that the null hypothesis is the one which contains the *equals* sign.

(3) In order to calculate the test statistic we need the data from the sample (to give us $\bar{x}$, s and n) and the null hypothesis (to give us a hypothesized value for μ). If the null hypothesis were true we would expect the value of the test statistic to be small. Conversely, the larger the value of the test statistic the more confidently we can reject the null hypothesis.

(4) Critical values are obtained from statistical tables. For a *t*-test we use the *t*-table, but many other tables are available and some will be used later. Users of significance tests do *not* calculate their own critical values; they use published tables, the construction of which will be discussed in a later chapter. The critical values are used as yardsticks against which we measure the test statistic in order to reach a decision concerning the null hypothesis. The decision process is illustrated by Fig. 4.1.

(5) In step 5 of the significance testing procedure we have quoted *two* critical values, at the 5% and the 1% significance levels. Many scientists use only *one* significance level, whereas others use two or even three. In different branches of science different conventions are adopted. If only one significance level is used this should be chosen *before* the data are analysed. Furthermore, the choice should not be left to a statistician. The scientist

selects his level of significance to correspond with the *risk* he is prepared to take of rejecting a null hypothesis which is true. If the chief chemist wishes to run a 5% risk of concluding that 'The modification has been effective', when in fact no increase in yield has occurred, then he should use a 5% significance level. Many social scientists and many biologists prefer to use several significance levels and then report their findings as indicated in Fig. 4.1. The chief chemist, for example, could report 'A significant increase in yield was found ($p < 0.05$)'.

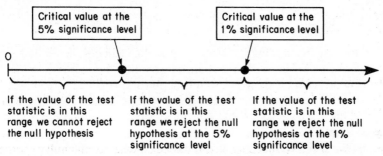

Figure 4.1   Comparing the test statistic with the critical values

(6) Significance tests can be classed as one-sided tests or two-sided tests. Each of the three *t*-tests we have carried out was a one-sided test because we were looking for a change in *one* direction. This one-sided search is indicated by the 'greater than' in the alternative hypothesis ($\mu > 70.30$). With a two-sided test the alternative hypothesis would state 'The mean yield of all batches after the modification is *not equal to* 70.30', (i.e. $\mu \neq 70.30$). We will return to this point in Chapter 5.

(7) In order to obtain critical values from the *t*-table we must use the appropriate number of degrees of freedom. For a one-sample *t*-test this will be 1 less than the number of observations used in the calculation of the standard deviation. When we carried out a significance test using the yield values from 50 batches we had 49 degrees of freedom and when we used the yield values from only 6 batches we had 5 degrees of freedom.

(8) In *almost all* significance tests (and certainly in *all* tests in this book) a large test statistic is an indication that either:

   (a) the null hypothesis is not true, or
   (b) the sample is not representative of the population referred to in the null hypothesis.

   If we have taken a random sample there is a good chance that it *will* be representative of the population, in that the sample mean ($\bar{x}$) will be close to the population mean (μ). When we get a large test statistic, therefore, we reject the null hypothesis.

Note that a small test statistic does not inspire us to *accept* the null hypothesis. At the decision step we either reject the null hypothesis or we fail to reject the null hypothesis. It would be illogical to accept the null hypothesis since the theory underlying the significance test is based on the assumption that the null hypothesis is true.

4.4   What is the new mean yield?

The chief chemist has concluded that the mean yield is now greater than 70.30. Having rejected the null hypothesis at the 5% level of significance he is quite confident that his conclusion will be verified as further batches are produced. He feels quite sure that the mean yield of future batches will be greater than 70.30.

In reporting his findings, however, the chief chemist feels that a more positive conclusion is needed. When informed of the increase in yield, the production director is sure to ask 'If the modification has been effective, what is the new mean level of yield?' The chief chemist would like, therefore, to include in his report an estimate of the mean yield that could be expected from future batches.

The 50 batches produced since the modification have given a mean yield of 71.23 (i.e. $\bar{x} = 71.23$). Should the chemist simply pronounce then that the mean yield of future batches will be 71.23 (i.e. $\mu = 71.23$)? If he does so he will almost certainly be wrong. Common sense suggests that if he had taken a different sample he would have got a different mean. If, for example, he had taken only the first 20 batches after the modification his sample mean would have been 71.17. The chief chemist surmises that if the sample mean varies from sample to sample it is quite possible that *none* of these sample means will be exactly equal to the population mean. To state that $\mu = 71.23$ simply because his sample has given $\bar{x} = 71.23$, would be unwise. (For a more detailed discussion of how the sample mean can be expected to vary from sample to sample see Appendix C.)

One solution to this problem is to quote a *range of values* in which we think the population mean must lie. Thus the chief chemist could report that the mean yield of future batches will lie in the interval 71.23 ± 0.15 (i.e. 71.08 to 71.38), for example. To quote an interval of values in which we expect the population mean to lie, does seem to be an improvement on the single value estimate. It is still open to question, however, and the chief chemist might well be asked 'How certain are you that the mean yield of future batches will be between 71.08 and 71.38?'

To overcome this difficulty we quote an interval *and* a confidence level in what is known as a *confidence interval*. To calculate such an interval we use the formula:

> A confidence interval for the population mean (μ) is given by $\bar{x} \pm ts/\sqrt{n}$

To perform this calculation we obtain $\bar{x}$, s and n from the sample data; t is a critical value from the *t*-table for a *two-sided* test using the appropriate degrees of freedom. Using 49 degrees of freedom the critical values are 2.01 at the 5% significance level and 2.68 at the 1% level. We can use these critical values to calculate a 95% confidence interval and a 99% confidence interval as follows.

A 95% confidence interval for μ

$$= 71.23 \pm 2.01 \, (3.251)/\sqrt{50}$$

$$= 71.23 \pm 0.92$$

$$= 70.31 \text{ to } 72.15$$

A 99% confidence interval for μ
$$= 71.23 \pm 2.68 \, (3.251)/\sqrt{50}$$

$$= 71.23 \pm 1.23$$

$$= 70.00 \text{ to } 72.46$$

Having carried out these calculations it is tempting to conclude, as many books do, that 'We can be 95% confident that the population mean lies between 70.31 and 72.15'. On reflection, however, the statement appears to have little meaning. The author does not judge himself competent to explain what is meant by the phrase 'We can be 95% confident', and he would certainly not wish to dictate to the reader how confident he or she should feel. However, several statements can be made which have a sound statistical basis. For example:

> If the experiment, with 50 batches, were repeated many, many times and a 95% confidence interval calculated on each occasion, then 95% of the confidence intervals would contain the population mean.

If the phrase 'We can be 95% confident' is simply an abbreviation of the above statement then it is reasonable to suggest that 'We can be 95% confident that the mean yield of future batches will lie between 70.31 and 72.15'. If we wish to be even more confident that our interval includes the population mean then we must quote a wider interval and we find that the 99% confidence interval is indeed wider, from 70.00 to 72.46.

When quoting a confidence interval we are making use of statistical inference just as we did when carrying out a significance test. We are again arguing from the sample to the population. A different sample would almost certainly give us a different confidence interval. If we had monitored only the first ten batches produced after the modifications we would have found $\bar{x} = 70.75$ and $s = 2.966$. Using $n = 10$ and $t = 2.26$ (9 degrees of freedom) we obtain a 95% confidence interval which extends from 68.63 to 72.87. This is very much wider than the interval obtained when we used the data from all 50 batches.

A larger sample can be expected to give us more information and we would expect a narrower confidence interval from a larger sample. This is reflected in the $\sqrt{n}$ divisor in the formula. It is interesting to compare the confidence intervals we would have obtained if we had examined only the first five batches or the first ten batches, etc., after the modifications had been incorporated. These 95% confidence intervals for the mean yield of future batches are given in Table 4.2 and are illustrated in Fig. 4.2.

Table 4.2   95% confidence intervals for mean yield of future batches

Number of batches	Sample mean	Sample SD		Confidence interval	
n	$\bar{x}$	*s*	*t*		
5	70.92	2.443	2.78	70.92 ± 3.04	67.88 to 73.96
10	70.75	2.966	2.26	70.75 ± 2.12	68.63 to 72.87
20	71.15	3.261	2.10	71.15 ± 1.53	69.62 to 72.68
30	71.26	3.417	2.04	71.26 ± 1.27	69.99 to 72.53
40	71.19	3.280	2.02	71.19 ± 1.05	70.14 to 72.24
50	71.23	3.251	2.01	71.23 ± 0.92	70.31 to 72.15

Concerning the confidence intervals in Table 4.2 and Fig. 4.2, several points are worthy of note:

(1) There is some indication that the sample mean and the sample standard deviation increase as the sample size (*n*) increases. This is simply due to chance and is *not* an indication of a general rule.
(2) The *t* value decreases as the sample size (*n*) increases. This is certainly not due to chance but is due to the greater reliability of the estimate of population standard deviation as the sample size increases. It would lead us to expect a narrower interval from a larger sample. (The ratio of the largest

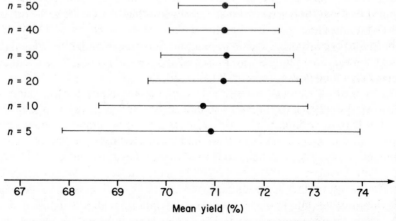

Figure 4.2   95% confidence intervals for mean yield of future batches

t value (2.78) to the smallest (2.01) is not very large and will not have a great effect on the width of the confidence interval. If, however, we calculated a 95% confidence interval using only the first *two* batches the *t* value would be 12.71 and the interval would extend from 56.12 to 84.08.)

(3) The width of the confidence intervals gets narrower as the sample size gets larger. This is largely due to the $\sqrt{n}$ divisor in the equation. Because of the square root this will not be a linear relationship and we would need, for example, to *quadruple* the sample size in order to *halve* the width of the confidence interval.

4.5 Is such a large sample really necessary?

Perhaps you get the impression from Fig. 4.2 that there is little to be gained by increasing the sample size above 30. It can be predicted in advance that a sample size of 120 would be needed to halve the width of the confidence interval that would be expected with a sample size of 30. Is it worth the extra effort and the increased cost to get the better estimate?

In carrying out an investigation we are, after all, *buying* information. The cost of the investigation may well be directly proportional to the sample size but the amount of information may only be proportional to the square root of the sample size. This would imply that there is a 'best' sample size.

Alternatively, we might wish to carry out an investigation in order to decide whether or not a process improvement has been achieved. The larger the improvement we are looking for the easier it should be to find, if it has actually been achieved. It is desirable that we should be able to decide *in advance* how large a sample is needed in order to detect a change of a specified magnitude.

Perhaps it would have been convenient for the chief chemist if he could have calculated in advance just how many batches he would need to examine in order to detect a mean yield increase of 1 unit (i.e. from 70.30 to 71.30). Unfortunately, such calculations are only possible if we have some information about the *variability* of the process. It is, of course, the batch to batch variability which prevents us from detecting immediately any change in mean yield which might occur.

The chief chemist is in a good position to calculate the number of batches he would need to examine in order to detect a *future* change because he now has a measure of process variability. This is in the form of a standard deviation calculated from the yield values of the 50 batches ($s = 3.251$). It can be substituted into the formula given below.

> The sample size (n) needed to detect a change (c) in population mean, using a one-sample *t*-test, is given by:
>
> $$n = (2ts/c)^2$$
>
> where s is the standard deviation of a previous sample of size m, and t has ($m - 1$) degrees of freedom.

Note that the degrees of freedom used to obtain the *t* value will be 1 less than the number of observations used to calculate the *existing* standard deviation. If we are looking for a change in one direction (e.g. an increase) we use a one-sided *t* value so with 49 degrees of freedom the *t* values are:

1.68 at the 5% significance level;
2.41 at the 1% significance level.

Suppose we wish to detect an increase in mean yield of 1 unit (i.e. $c = 1$); then:

For 95% confidence

$$n = [2(1.68)\ (3.251)/1]^2 = 119.3$$

For 99% confidence

$$n = [2(2.41)\ (3.251)/1]^2 = 245.5$$

These results tell us that we would need a sample size of 120 or more (rounding up 119.3) in order to be 95% confident of detecting (by means of a *t*-test) an increase in mean yield of 1 unit. If we wish to be 99% confident of detecting such a change in mean we would need to examine at least 246 batches.

The sample size needed to detect a change in mean will depend on the *size* of change we wish to detect. A smaller sample size will suffice if it is a larger change that we are seeking. Substituting $c = 2$ into the formula gives $n = 29.8$, with a 5% *t* value. Thus we need inspect only 30 batches to be 95% confident of detecting an increase in mean yield of 2 units (i.e. from 70.3 to 72.3).

If we are interested in *estimating* a population mean to within a given tolerance ($\pm c$) we can use the formula given below.

> The size of sample needed to estimate the population mean to within $\pm c$, using a confidence interval is given by:
>
> $$n = (ts/c)^2$$

The value of *t* is dependent on the degrees of freedom and also whether a one-sided or two-sided confidence interval is of interest.

4.6   Statistical significance and practical importance

Perhaps the time is ripe to summarize the statistical activity of the chief chemist. Firstly, he carried out a significance test using the yield values of the first 50 batches produced after the modifications had been incorporated. He concluded that the mean yield had increased and was now greater than the old mean of 70.3. Secondly, he calculated a confidence interval for the mean yield of the modified plant. He was able to report with 95% confidence that the mean yield is now between 70.31 and 72.15. Finally, he calculated the size of sample that would be needed to detect a specified change in mean yield. Using the standard deviation of yield values from the 50 batches he was able to show that a sample of 120 batches would be needed to detect a mean increase of 1 unit in yield with 95% confidence.

What benefits has the chief chemist gained by his use of statistical techniques? A significance test has enabled him to demonstrate that the apparent increase in yield indicated by the first 50 batches is very likely a reflection of a real and permanent increase. That does *not* imply that this increase is of practical importance or has great financial value. The confidence interval would indicate that the increase in mean yield is around 1 unit but it could be as much as 2 units. The confidence interval, like the significance test, is indicating statistical significance and *not* practical importance.

The statistician has no desire to tell the chemist what constitutes an *important* change. The statistician would certainly not wish the chemist to equate statistical significance with practical importance. Only the chemist can decide what is important and he must take into account *many* factors including the possible consequences of his decision.

For example, Indichem manufacture two pigments which are very *similar* in chemical composition and method of production. In terms of profitability to the company, however, they are very *different*. Pigment A is in great demand and is manufactured in an uninterrupted stream of batches in one particular vessel. Pigment B, on the other hand, is in low demand and only one batch is manufactured each year. Production costs of pigment A are dominated by the cost of raw materials and a 1% increase in yield would be worthwhile. Production costs of pigment B are much more dependent on the costs of the labour intensive cleaning and setting-up, so that a 1% increase in yield would be considered trivial.

The chemist must decide, then, whether or not $x\%$ change is important. He must *not*, however, equate percentage change with *significant* change. When assessing the results of an investigation it is important to ask *two* questions:

(a) Could the observed change be simply due to chance?
(b) If the change is not due to chance, is it large enough to be important?

Significance testing helps us to answer the first question, but we must bear in mind that an observed change of $x\%$ may be significant if it is based on a *large* sample but not significant if based on a *small* sample. A significance test automatically takes into account the sample size and the inherent variability which is present in the data, whereas a simple statement of percentage increase does not.

4.7 Summary

This is a very important chapter. In it we have introduced the fundamentals of significance testing and estimation. These ideas will serve as a foundation for later chapters and should be consolidated by working through the problems before proceeding further. When attempting the problems you would be well advised to follow the six-step procedure which will be used again and again in later chapters.

There are some difficult concepts underlying significance testing and if you worry about details which have not been explained you may lose sight of why a chemist or technologist would wish to do a significance test. Try to remember that the purpose of a significance test is to reach a decision about a *population* using data obtained from a *sample*.

Problems

(1) A polymer manufacturer specifies that his produce has a mean intrinsic viscosity greater than 6.2. A customer samples a consignment and obtains the following values:

   6.21   6.32   6.18   6.34   6.26   6.31

Do the results confirm the manufacturer's claim?

(2) A new analytical method has been developed by a laboratory for the determination of chloride in river samples. To evaluate the accuracy of the method eight determinations were made on a standard sample which contained 50 mg/litre. The determinations were:

   49.2   51.1   49.6   48.7   50.6   49.9   48.1   49.6

(a) Carry out a significance test to determine whether the method is biased.
(b) Calculate a 95% confidence interval for the population mean chloride content. Using this interval what is the maximum bias of the method?
(c) What is the population under investigation?

(3) A recent modification has been introduced into a process to lower the impurity. Previously the impurity level had a mean of 5.4. The first ten batches with the modification gave the following results:

   4.0   4.6   4.4   2.5   4.8   5.9   3.0   6.1   5.3   4.4

(a) Has there been a decrease in mean impurity level?
(b) Obtain a 95% confidence interval for the mean impurity level.
(c) How many results are needed in order to be 95% confident of estimating the mean impurity to within ± 0.15?

5

Testing and estimation: two samples

5.1 Introduction

The previous chapter was centred around the *one-sample t-test*. We made use of this particular type of significance test to decide whether or not the mean yield of a production process had increased as a result of carrying out a modification. We showed that the mean yield of a sample of 50 batches was significantly greater than a standard value (70.3) which had previously been established.

In this chapter we will progress to the *two-sample t-test*. Once again we will be using the six-step procedure to reach a decision about a null hypothesis so the t-test itself should present no problems. There are, however, two complications which cannot be ignored:

(a) There are *two* types of two-sample t-test and we must be careful to select the appropriate type.
(b) Before we can use a t-test to compare two sample means we must *first* compare the sample variances using a different type of significance test which is known as the F-test.

Because the F-test is a prerequisite of the two-sample t-test we will examine the former before the latter. First of all, therefore, we will explore a situation in which the F-test is very meaningful in its own right. This will necessitate leaving the production process and moving into the analytical laboratory. There we will make use of the F-test to compare the precision of two alternative test methods.

5.2 Comparing the precision of two test methods

In previous chapters we examined the impurity determinations made on the 50 batches of digozo blue. Table 2.1 contains *one* determination for each batch. It was not mentioned earlier but each of these determinations is the mean of five measurements. Because of the inherent variability of the analytical test method the specification calls for the measurement to be repeated five times and then the five measurements to be averaged to obtain the single determination of impurity for the batch. For example the impurity determination for the 50th batch (1.29) was calculated by averaging:

   0.87   1.12   1.66   1.20   and   1.60

The chief chemist is not entirely satisfied with the test method as it stands. For some time he has suspected that operators are not recording faithfully the measurements obtained. He feels that the temptation to 'bring in' the extreme measurements is too strong for some operators when confronted with five measurements. He would like to reduce the number of repeat measurements to three or even two, but he could only afford to do so if it were possible to reduce the inherent variability of the test method.

   The chief chemist decides to modify the test method in an attempt to improve its precision (i.e. reduce its variability). Unfortunately this change will increase the cost of each measurement so he would not wish to adopt the new method until it has been shown to be less variable than the method now in use. As a preliminary investigation he asks an operator to make six measurements with the new method on batch number 50. The new impurity measurements are:

   1.29   1.72   1.20   1.25   1.65   1.48

How do these six measurements compare with those made earlier using the old analytical method? The mean and standard deviation of both sets of measurements are given in Table 5.1, whilst Fig. 5.1 offers a pictorial comparison which is perhaps more useful.

Table 5.1   Determination of impurity by two analytical methods

	Mean	*Standard deviation*
New method	1.432	0.2190
Old method	1.290	0.3341

   Summarized in Table 5.1 and Fig. 5.1 are two *samples* of measurements. By the old method we have one sample of five results and by the new method we have one sample of six results. If we continued testing indefinitely we would

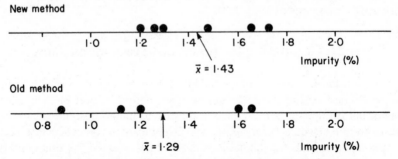

Figure 5.1   Determination of impurity by two analytical methods

obtain two *populations* of results. Obviously it is not possible to continue testing forever but it is important to question the nature of the two populations when comparing the two analytical methods. To simplify the discussion and to distinguish clearly between populations and samples we will use the internationally accepted symbols given in Table 5.2.

Table 5.2   Symbols used in the *F*-test and the two-sample *t*-test

	Sample		Population	
	Mean	*SD*	*Mean*	*SD*
New method	$\bar{x}_1$	s_1	μ_1	σ_1
Old method	$\bar{x}_2$	s_2	μ_2	σ_2

Clearly the sample means are not equal ($\bar{x}_1 = 1.43$ and $\bar{x}_2 = 1.29$) but is it possible that the population means are equal? Later we will use a two-sample *t*-test to answer this question and in this test the null hypothesis will be $\mu_1 = \mu_2$. Before we can carry out the *t*-test, however, we must concern ourselves with the *variability* of the impurity measurements.

It is equally clear that the two sample standard deviations are not equal ($s_1 = 0.2190$ and $s_2 = 0.3341$) but is it possible that the population standard deviations are equal? To answer this question we can use an *F*-test in which the null hypothesis will be $\sigma_1^2 = \sigma_2^2$. (The *F*-test is based on variances not standard deviations, but the two are very closely related, of course.) If this null hypothesis is true the two analytical methods are equally variable; in other words they have equal precision. We have found that the sample standard deviations are *not* equal but it is possible that the two methods would be revealed to be equally variable if testing had been continued. When carrying out an *F*-test the test statistic is calculated by means of the following formula:

> *F-test*
> Test statistic = larger variance/smaller variance

The *F*-test proceeds as follows:

Null hypothesis – The two analytical methods have equal precision ($\sigma_1^2 = \sigma_2^2$).

Alternative hypothesis – The new method is more precise than the old method ($\sigma_1^2 < \sigma_2^2$).

Test statistic $= s_2^2/s_1^2$

$$= (0.3341)^2/(0.2190)^2$$

$$= 2.33$$

Critical values – From the *F*-table (Statistical Table C), using 4 degrees of freedom for the larger variance and 5 degrees of freedom for the smaller variance (one-sided test), the critical values are:

5.19 at the 5% significance level
11.39 at the 1% significance level.

Decision – As the test statistic is less than both critical values we cannot reject the null hypothesis.

Conclusion – We are unable to conclude that the new method is more precise than the old method.

It appears then that the apparently smaller variability of the new method of test could simply be due to chance, and it would be unwise to conclude that the new method had superior precision. If the chief chemist is not satisfied with this conclusion he can, of course, carry out a more extensive investigation. He could embark upon a full scale precision experiment involving several laboratories, many operators and more than one batch. Such inter-laboratory trials are discussed in a separate volume, *Statistics for Analytical Chemists* (Caulcutt and Boddy, 1983).

The *F*-test we have just carried out was a *one-sided* significance test. The one-sided test was appropriate as we were looking for a change in *one* direction, i.e. an improvement. The chief chemist does not seek a new method which is different, he seeks a method which is *better*. This point is emphasized by the presence of $<$ in the alternative hypothesis, rather than $\neq$ which would have indicated the need for a two-sided test.

Note that we would not have calculated a test statistic if the sample variance of the new method (s_1^2) had been greater than the sample variance of the old method (s_2^2). Obviously the discovery that $s_1^2 > s_2^2$ does *not* constitute support for the alternative hypothesis that $\sigma_1^2 < \sigma_2^2$.

> When carrying out a *one-sided* significance test we only calculate a test statistic if the sample data appear to support the alternative hypothesis. If the data appear to contradict the alternative hypothesis then we certainly cannot reject the null hypothesis.

Consider, for example, the one-sided, one-sample *t*-test in Chapter 4. When carrying out this test it would have been very foolish to conclude that the mean yield had *increased* (i.e. $\mu > 70.30$) if the mean yield of the sample had been *less than* the old mean yield (i.e. $\bar{x} < 70.30$).

5.3   Comparing the bias of two test methods

When comparing two analytical test methods we must never forget that the more precise method is *not necessarily* the more accurate. Precision is concerned solely with consistency or self-agreement. Whilst precision is highly desirable it does not guarantee accuracy. It is possible, for example, that a third method of measuring impurity might have given the following measurements when applied repeatedly to batch number 50:

0.81   0.81   0.81   0.81

This third method appears to have very high precision but it may not be accurate. If the *true* impurity is 1.30, say, then the first method is superior even though it is less precise. We would say that method three was *biased* whereas the first method was unbiased. If we are to declare an analytical method to be *accurate* then we must be satisfied that it has adequate precision *and* that it is unbiased.

In practice it may be easy to see that a test method lacks precision but to detect bias we need to know the true value of that which is being measured. For the chief chemist to establish that either test method was unbiased he would need to know the *true* impurity of batch number 50. Unfortunately he does not have this information. We can state categorically, then, that with the impurity measurements he has obtained so far, he is in no position to conclude that:

(a) either the old method is unbiased
(b) or the new method is unbiased.

Despite this limitation, which is imposed by his not knowing the true impurity, the chief chemist can compare the average performance of the two methods. If the two sample means ($\bar{x}_1 = 1.43$ and $\bar{x}_2 = 1.29$) are significantly different this might give some indication that one or other of the methods is biased. To compare the two means we would use a significance test which is known as a '*t*-test for two independent samples' and the test statistic is calculated using the following formula:

> *t*-test for two independent samples
>
> $$\text{Test statistic} = \frac{|\bar{x}_1 - \bar{x}_2|}{s\sqrt{\left(\dfrac{1}{n_1} + \dfrac{1}{n_2}\right)}}$$
>
> where
>
> $$s = \sqrt{\{[(n_1 - 1)s_1^2 + (n_2 - 1)s_2^2]/[(n_1 - 1) + (n_2 - 1)]\}}$$
>
> and s_1^2 does *not* differ significantly from s_2^2.

The *t*-test for two independent samples makes use of our six-step procedure as follows:

Null hypothesis – The two analytical methods would give the same mean measurement in the long run (i.e. $\mu_1 = \mu_2$).

Alternative hypothesis – The two analytical methods would not give the same mean measurement in the long run (i.e. $\mu_1 \neq \mu_2$).

$$s = \sqrt{\{[(n_1 - 1)s_1{}^2 + (n_2 - 1)s_2{}^2]/[(n_1 - 1) + (n_2 - 1)]\}}$$
$$= \sqrt{\{[(6 - 1)(0.2190)^2 + (5 - 1)(0.3341)^2]/[(6 - 1) + (5 - 1)]\}}$$
$$= 0.2761$$

$$\text{Test statistic} = \frac{|\bar{x}_1 - \bar{x}_2|}{s\sqrt{\left(\dfrac{1}{n_1} + \dfrac{1}{n_2}\right)}}$$

$$= \frac{|1.43 - 1.29|}{0.2761\sqrt{\left(\dfrac{1}{6} + \dfrac{1}{5}\right)}}$$

$$= 0.83$$

Critical values – From the t-table (Statistical Table B) using 9 degrees of freedom [i.e. $(n_1 - 1) + (n_2 - 1)$] for a two-sided test:

2.26 at the 5% significance level
3.25 at the 1% significance level.

Decision – As the test statistic is less than both critical values we cannot reject the null hypothesis.

Conclusion – We are unable to conclude that there is any significant difference between the means of the results from the two analytical methods.

It could be, then, that in the long run neither of the two methods will be superior to the other in terms of precision or bias. As the new method is more costly to use it would be unwise to introduce it on the strength of the very minimal evidence gathered to date.

You will have noticed that in carrying out the two-sample t-test we chose to do a *two-sided* test. This was appropriate because we were attempting to answer the question 'Do the two methods give significantly *different* results on average?' We were looking for a difference in *either* direction. Previously when carrying out the F-test we chose to do a one-sided test because we were looking for a change in one particular direction. We were seeking a *decrease* in variability of the new method with respect to the old method.

The 't-test for two independent samples' is probably the most widely used of all significance tests. Because of its importance we will explore a second application of this test before we examine the second type of two-sample t-test.

5.4 Does the product deteriorate in storage?

The chief chemist suspects that the brightness of the digozo blue pigment decreases whilst the product is in storage. His opinion on this matter has been influenced by recent complaints from Asian customers about the dullness of pigment which was manufactured some years ago. In order to investigate the occurrence of this deterioration the chief chemist has measured the brightness of the pigment in ten drums of digozo blue which have been in the warehouse for many months. It is known that these drums are not all from the same batch and it is thought that they are the residue from at least three different batches. The brightness determinations are:

2   1   3   0   1   3   1   1   1   0

As a basis for comparison the chief chemist measures the brightness of the pigment produced more recently. This is taken from ten drums of digozo blue, five drums from each of the last two batches. The brightness determinations are:

2   2   4   1   2   3   3   2   1   2

You may well be struck by the coarseness of the measurements. Brightness is measured on a scale which ranges from -7 to $+7$ with more positive numbers corresponding to greater brightness. Only five points of the scale (0, 1, 2, 3 and 4) are covered by the data, however. If each measurement had a tolerance of ± 1 there would appear to be little information on which to base a decision. You may have further reservations concerning the 'unit of measurement'. Is the difference between 1 and 2 on the brightness scale really equal to the difference between 3 and 4? The colour physicist who established the scale is confident that the answer is 'yes', and that a genuine unit of measurement does exist. If this were *not* so, the data would be unsuitable for the use of an *F*-test or a *t*-test. Many statisticians would express concern about the use of these tests because of the coarseness of the data. Unfortunately the alternative significance tests which one might adopt in this situation suffer from two defects:

(a) they are less powerful than the *t*-test or *F*-test;
(b) they are not within the repertoire of the majority of scientists (nor are they included in this book).

It is certainly true to say that the ubiquitous *t*-test is occasionally used in situations where it is not strictly valid and we will now make use of the *t*-test to compare the brightness of the two samples.

The two sets of measurements are summarized in Table 5.3. We see in this table that the mean brightness of the old pigment (1.3) is less than the mean brightness of the new pigment (2.2). To see if the difference between the sample means is significant we will carry out a *t*-test for two independent samples. Before we do so, however, we must check that brightness

Table 5.3   Brightness of old and new pigment

	Number of drums	Brightness	
		Mean	SD
Old pigment	$n_1 = 10$	$\bar{x}_1 = 1.3$	$s_1 = 1.059$
New pigment	$n_2 = 10$	$\bar{x}_2 = 2.2$	$s_2 = 0.919$

measurements are equally variable whether made on old pigment or new pigment. For this purpose we carry out an *F*-test as follows:

Null hypothesis: $\sigma_1^2 = \sigma_2^2$ (i.e. variability of brightness is the same for new and old pigment).

Alternative hypothesis: $\sigma_1^2 = \sigma_2^2$.

Test statistic $= (1.059)^2/(0.919)^2$

$= 1.33$

Critical values – From the *F*-table using 9 degrees of freedom for a two-sided test:

4.03 at the 5% significance level
6.54 at the 1% significance level.

Decision – We cannot reject the null hypothesis.

Conclusion – We cannot conclude that the two population variances are unequal. Thus we provisionally assume that variability of brightness is the same for old and new pigment.

If we had rejected the null hypothesis when carrying out the *F*-test we would have been unable to proceed with the *t*-test for two independent samples. As we did not reject the null hypothesis we *can* proceed with the *t*-test on the assumption that the population variances are equal. The *t*-test proceeds as follows:

Null hypothesis: $\mu_1 = \mu_2$ (i.e. the mean brightness of old pigment is equal to the mean brightness of new pigment).

Alternative hypothesis: $\mu_1 < \mu_2$ (i.e. the mean brightness of old pigment is less than the mean brightness of new pigment)

Combined standard deviation:

$s = \sqrt{\{[(n_1 - 1)s_1^2 + (n_2 - 1)s_2^2]/[(n_1 - 1) + (n_2 - 1)]\}}$

$= \sqrt{\{[9(1.059)^2 + 9(0.919)^2]/[9 + 9]\}}$

$= 0.992$

$$\text{Test statistic} = \frac{|\bar{x}_1 - \bar{x}_2|}{s\sqrt{\left(\dfrac{1}{n_1} + \dfrac{1}{n_2}\right)}} = \frac{|1.3 - 2.2|}{0.992\sqrt{\left(\dfrac{1}{10} + \dfrac{1}{10}\right)}}$$

$$= 2.02$$

Critical values – From the *t*-table using 18 degrees of freedom for a one-sided test:

1.73 at the 5% significance level
2.55 at the 1% significance level.

Decision – We reject the null hypothesis at the 5% significance level.

Conclusion – We conclude that the mean brightness of old pigment is less than the mean brightness of new pigment.

The evidence furnished by the 20 drums of pigment leads us to the conclusion that the *sample* means are significantly different. It is reasonable therefore to ask 'How large is the difference between the *population* means?' This question can only be answered by using the sample means as estimates of the population means and as usual we express the uncertainty of our estimates in a confidence interval.

A confidence interval for the difference between two population means is given by:

$$|\bar{x}_1 - \bar{x}_2| \pm ts\sqrt{\left(\frac{1}{n_1} + \frac{1}{n_2}\right)}$$

where s is the combined standard deviation:

$$s = \sqrt{\{[(n_1 - 1)s_1^2 + (n_2 - 1)s_2^2]/[(n_1 - 1) + (n_2 - 1)]\}}$$

To calculate a 95% confidence interval we would use a two-sided *t* value and a 5% significance level. With ten drums in each sample we have 18 degrees of freedom and the *t* value is 2.10, giving the following confidence interval:

$$|1.3 - 2.2| \pm 2.10\,(0.992)\sqrt{\left(\frac{1}{10} + \frac{1}{10}\right)}$$

$$= 0.9 \pm 0.94$$

$$= -0.04 \text{ to } 1.84$$

This result is telling us that we can be 95% certain that the reduction in brightness during storage will be less than 1.84 and greater than −0.04 (i.e. a slight gain in brightness). You might be surprised that this confidence interval should include zero when the *t*-test has already told us that the difference between the two population means is greater than zero. The reason why the confidence interval and the *t*-test are not in complete agreement is because in one case we used a two-sided *t* value whilst in the other we used a one-sided *t* value. It is perfectly reasonable that we should carry out a one-sided *t*-test because we were looking for a *reduction* in brightness which is a change in *one* direction. (Had we done a two-sided test, incidentally, we would *not* have rejected the null hypothesis.) On the other hand it is common practice to use a two-sided *t* value when calculating a confidence interval. The confidence intervals we have calculated in this and earlier chapters should strictly speaking be called 'two-sided confidence intervals' to distinguish them from the much less common 'one-sided confidence intervals'. The latter would be calculated using a one-sided *t* value.

In our particular problem then we could use the one-sided *t* value (1.73) that we used in the *t*-test, to calculate *either* a lower confidence limit *or* an upper confidence limit as follows:

Lower limit for the difference between the population means

$$= |\bar{x}_1 - \bar{x}_2| - ts \sqrt{\left(\frac{1}{n_1} + \frac{1}{n_2}\right)}$$

$$= 0.9 - 1.73 \, (0.992) \sqrt{\left(\frac{1}{10} + \frac{1}{10}\right)}$$

$$= 0.13$$

Upper limit for the difference between the population means

$$= |\bar{x}_1 - \bar{x}_2| + ts \sqrt{\left(\frac{1}{n_1} + \frac{1}{n_2}\right)}$$

$$= 0.9 + 1.73 \, (0.992) \sqrt{\left(\frac{1}{10} + \frac{1}{10}\right)}$$

$$= 1.67$$

Thus we can state that we are 95% confident that the reduction in brightness is greater than 0.13 *or* we can state that we are 95% confident that the reduction in brightness is less than 1.67 but we *cannot* make the two statements simultaneously. We could, however, quote 0.13 and 1.67 as the limits of a *90% confidence interval.*

Whether we are calculating a one-sided confidence interval or the much more usual two-sided confidence interval we would expect the width of the interval to depend upon the sample size. In the last chapter we used the formula:

$$n = \left(\frac{2ts}{c}\right)^2$$

to calculate the sample size needed to detect a change c in the population mean with a certain level of confidence. That was in the simple situation where we intended to take *one* sample from one population.

In the more complex situation where we wish to detect or estimate a difference between *two* population means by taking a sample from each population, we can make use of the formulae:

> The smallest sample sizes needed to detect a difference c between *two* population means, using a t-test for two independent samples, are given by:
>
> $$n_1 = n_2 = 2 \left(\frac{2ts}{c} \right)^2$$
>
> where s is the standard deviation of a previous sample of size m, and t has $(m-1)$ degrees of freedom.

> The smallest sample sizes needed to estimate the difference between *two* population means to within $\pm c$ are given by:
>
> $$n_1 = n_2 = 2 \left(\frac{ts}{c} \right)^2$$

Thus we can calculate that sample sizes of 24 would be needed to be 95% confident of detecting a decrease in brightness of 1 unit, by means of a t-test for two independent samples. To obtain this sample size we let $t = 1.73$ (one-sided with 18 degrees of freedom), $s = 0.992$ and $c = 1$.

5.5   A much better experiment

The investigation carried out by the chief chemist is far from ideal. You will recall that he wished to determine whether or not the brightness of digozo blue pigment decreased during an extended period of storage. He compared the mean brightness of ten drums of pigment that had been recently manufactured, with the mean brightness of ten drums which were produced many months ago. By means of a two-sample t-test he has demonstrated that the two sample means are significantly different. Before the chief chemist concludes that brightness does decrease during storage there are several questions he should ask:

(a) Were the ten drums of new pigment representative of *all* drums of digozo blue produced recently?
(b) Were the ten drums of old pigment representative of *all* drums produced at that time?
(c) Was the mean brightness of all old drums the same, at the time they were produced, as the present mean?

It is very doubtful if the chief chemist is in a position to answer all of these questions and we must, therefore, regard his investigation with some suspicion. Perhaps he would be wise to put his results on one side and carry out an experiment which is more suited to detecting a change in brightness with age. The chief chemist would be able to speak with more authority if he carried out a *longitudinal* investigation. This would involve observing the *same* drums of pigment on two (or more) occasions. He could, for example, measure the brightness of ten drums of pigment selected at random from recently produced batches, and then repeat these measurements on the *same* ten drums in two years time. This experiment would be in contrast with the *cross-sectional* investigation that he has already carried out, in which two *different* samples are measured at the *same* time.

Suppose that the chief chemist had chosen to carry out a longitudinal experiment and had obtained the results of Table 5.4. We see in this table that the mean brightness has decreased from 2.2 to 1.3 during the period of storage. These figures may ring a bell because they occurred earlier. In fact the brightness measurements in Table 5.4 are *exactly* the same as those used in the *t*-test for two independent samples when we were analysing the cross-sectional experiment.

Table 5.4   Brightness of pigment, measured on two occasions

Drum	First occasion	Second occasion
A	2	2
B	2	1
C	4	3
D	1	0
E	2	1
F	3	3
G	3	1
H	2	1
I	1	1
J	2	0
Mean	2.2	1.3

Though the longitudinal experiment has yielded the same results it is *not* appropriate to use the same method of analysis. Instead of the '*t*-test for two independent samples' we will carry out a '*t*-test for two matched samples'. The latter test is often referred to as a 'paired comparison test' which is perhaps a better name as the word 'paired' indicates an important feature of the data in Table 5.4. Each brightness determination in the right hand column is *paired* with one of the determinations in the centre column. This pairing of the observations occurs, of course, because we have *two* measurements on each drum; a feature which could not be made use of if we carried out the independent samples test on the data in Table 5.4.

The paired comparison test is actually quite straightforward. We simply calculate the *decrease* in brightness for each of the ten drums, as in Table 5.5, and then carry out a one-sample *t*-test on these calculated values.

Table 5.5   Decrease in brightness

Drum	First occasion	Second occasion	Decrease d
A	2	2	0
B	2	1	1
C	4	3	1
D	1	0	1
E	2	1	1
F	3	3	0
G	3	1	2
H	2	1	1
I	1	1	0
J	2	0	2
Mean	–	–	$d = 0.9$
SD	–	–	$s_d = 0.738$

Null hypothesis – Mean decrease in brightness for *all* drums would be equal to zero (i.e. $\mu_d = 0$).

Alternative hypothesis – Mean decrease in brightness for *all* drums would be greater than zero ($\mu_d > 0$).

Test statistic $= \dfrac{|\bar{d} - \mu_d|}{s_d/\sqrt{n}}$

$\quad = \dfrac{|0.9 - 0.0|}{0.738/\sqrt{10}}$

$\quad = 3.86$

Critical values – From the *t*-table with 9 degrees of freedom for a one-sided test:

1.83 at the 5% significance level
2.82 at the 1% significance level.

Decision – We reject the null hypothesis at the 1% level of significance.

Conclusion – We conclude that the brightness of the pigment has decreased during the storage period.

Note that the test statistic (3.86) is much larger than that calculated when doing the *t*-test for two independent samples (2.02) using the same data. This arises because when we calculate differences in the paired comparison test we automatically eliminate the drum to drum variability. This becomes clear if we compare the standard deviations used in the calculation of the test statistic.

Paired comparison test: $\qquad\qquad\qquad s_d = 0.738$

t-test for two independent samples: $\quad s_1 = 1.059$ and $s_2 = 0.919$

The power of the paired comparison test will be illustrated more dramatically in the problems which follow.

5.6 Summary

In this chapter we have introduced three more significance tests to complement the one-sample t-test dealt with earlier. All four tests have followed the same six-step procedure which should now be very familiar.

We used the F-test to compare the precision of two test methods and also as a prerequisite to the 't-test for two independent samples'. The latter was used to check the statistical significance of the difference between two sample means, in order to decide whether or not the brightness of a pigment deteriorated during storage. Finally we used the paired comparison test for the same purpose, using data from a longitudinal experiment.

As your repertoire of significance tests grows you may become increasingly anxious that we are neglecting the *assumptions* which underlie these tests. Discussion of these assumptions will be reserved for the final chapter of Part One after we have examined more significance tests which are rather different from those studied so far.

Problems

(1) Two methods of assessing the moisture content of cement are available. On the same sample of cement the same operator makes a number of repeat determinations and obtains the following results (mg/cc):

| Method A | 60 | 68 | 65 | 69 | 63 | 67 | 63 | 64 | 66 | 61 |
| Method B | 65 | 64 | 62 | 62 | 64 | 64 | 65 | 66 | 64 | |

Is one method more repeatable than the other? (A method that is more repeatable will have less variability in repeat determinations.)

(2) Tertiary Oils have developed a new additive, PX 235, for their engine oil which they believe will decrease petrol consumption. To confirm this belief they carry out an experiment in which miles per gallon (m.p.g.) are recorded for different makes of car. Altogether twelve makes of car were chosen for the experiment and they were randomly split into two groups of six. One group used the oil with the additive, and the other group used oil without the additive. Results were:

| Oil with additive | 26.2 | 35.1 | 43.2 | 36.2 | 29.4 | 37.5 |
| Oil without additive | 37.1 | 42.2 | 26.9 | 30.2 | 33.1 | 30.9 |

(a) Are the population variances of the two oils significantly different?
(b) Has the new oil increased miles per gallon?
(c) How many cars would be needed in an investigation of this type in order to be 95% certain of detecting an improvement of 1.0 m.p.g.?

(d) What is the population referred to in (a)?

(3) The research manager of Tertiary Oils is most dissatisfied with the experimental design given in Problem 2. He recognizes that there is large variability between cars and that this can be removed by using a two-sample paired design. He therefore authorizes an experiment in which six cars are used. The following results are obtained:

Make of Car	A	B	C	D	E	F
Oil with additive	26.2	35.1	43.2	36.2	29.4	37.5
Oil without additive	24.6	34.0	41.5	35.9	29.1	35.3

(a) Has the additive significantly increased the miles per gallon?
(b) How many cars are needed to be 95% certain of detecting a difference of 1.0 m.p.g., when using this type of design?

----------------------------6----------------------------

Testing and estimation: proportions

6.1 Introduction

In the two previous chapters we have been primarily concerned with t-tests. We were looking for a significant difference between two sample means or between one sample mean and a specified value. The data used in these t-tests resulted from measuring such variables as yield, impurity and brightness.

In each of the situations we examined it was meaningful to calculate an average value. Clearly, it makes sense to speak of the mean yield, the mean impurity or the mean brightness of several batches of pigment. In many other situations we must deal with *qualitative* variables which cannot be averaged to produce a meaningful result. We might, for example, record whether a batch was made for export or for the home market. We might record that an employee was female or that she was located in the research department. 'Destination', 'sex' and 'location' are qualitative variables and when summarizing such variables it is useful to *count* and then to calculate *proportions* or percentages. We might speak of the proportion of batches made for export and to calculate this proportion we would need to *count* the total number of batches and the number made for export. We might also speak of the percentage of employees who are female or the proportion of staff located in the research department.

In this chapter we will examine a variety of significance tests which are applicable to proportions in particular and to counted data (i.e. frequencies) in general. As these tests are often used in situations which involve people, we will first examine an experiment concerned with the sensory evaluation of a product used in the home.

6.2 Does the additive affect the taste?

Duostayer Research Laboratories have been asked by their parent company to investigate the effectiveness of a new toothpaste additive, PX 235. The inclusion of PX 235 is expected to increase the protection given to the tooth enamel when the paste is used regularly, and the extent of this increased protection is to be estimated from the results of a major investigation carried out over a long period of time.

In addition to evaluating the effectiveness of the ingredient as a protector of enamel the laboratory is required to investigate the customer acceptibility of the toothpaste when PX 235 is included. This minor investigation will be carried out by the sensory evaluation department. Within this department Dr Murphy, a very experienced research chemist, has been given the responsibility for finding an answer to the question 'Does the inclusion of PX 235 affect the taste of the toothpaste?' Dr Murphy plans his investigation as follows:

A 1 kg batch of *normal* toothpaste and a 1 kg batch of *treated* toothpaste are each subdivided into 18 samples. Twelve members of a trained tasting panel are each confronted with three samples; the samples being set out as follows:

Assessor	Left	Centre	Right
1	N	T	T
2	T	N	T
3	T	T	N
4	N	N	T
5	N	T	N
6	T	N	N
7	N	T	T
8	T	N	T
9	T	T	N
10	N	N	T
11	N	T	N
12	T	N	N

(T = treated; N = normal)

The three samples presented to any assessor contain two which are identical and one which is different. Each assessor is asked to 'Select the sample which tastes different from the other two'. The twelve assessors are assigned at random to the twelve sets of samples which are laid out in accordance with the recommendations of draft BS 79/54233 DC: 'Sensory analysis – methodology – triangular test'.

Dr Murphy carries out the experiment and he finds that eight of the twelve assessors correctly select the odd one out, whilst the other four assessors fail to do so. What conclusion can he draw concerning the taste of the treated toothpaste?

Had he found that *all twelve* assessors correctly identified the odd one amongst the three samples, he would have been happy to conclude that the treated toothpaste *did* taste different. On the other hand if *none* of the assessors had been successful he would have readily concluded that the treated and untreated toothpastes were indistinguishable. Eight out of twelve does seem to Dr Murphy to be a borderline case. He realizes that it is quite possible for an assessor to *guess* correctly which of the three samples is the odd one.

In previous chapters we have used significance tests to help us reach decisions in borderline cases. The twelve assessors are, after all, only a *sample* taken from a population of assessors. Dr Murphy would really like to know what the result of his experiment would have been if the whole population of assessors had taken part.

To carry out a significance test we will need critical values. Draft British Standard 79/54233 DC contains a table of critical values which can be used as a yardstick by anyone in Dr Murphy's position. A simplified version of this table is included with the other statistical tables (Table D). The significance test proceeds as follows:

Null hypothesis – The treated and the untreated toothpastes have the *same* taste.

Alternative hypothesis – The treated and the untreated toothpastes do not have the same taste.

Test statistic = 8 (the number of assessors who were correct).

Critical values – From the triangular test table (Table D), for an experiment with twelve assessors:

7½ at the 5% significance level
8½ at the 1% significance level.

Decision – Because the test statistic is greater than the lower critical value we reject the null hypothesis at the 5% significance level.

Conclusion – We conclude that it *is* possible for assessors to distinguish between the treated and the untreated toothpaste.

Though this significance test has followed the same six-step procedure which served us well in the *t*-test and the *F*-test, it does differ in some respects from these more common tests. Perhaps the only fundamental difference is that in the triangular test we are dealing with a *discrete* variable. This discreteness is evident in the test statistic which can only be a whole number.

6.3   Where do statistical tables come from?

Tables of critical values must appear rather mysterious to many who use them, because they are rarely accompanied by an explanation which is comprehensible to anyone who is not a statistician. Fortunately the table of critical values for the triangular test is based on *relatively* simple statistical theory.

You will note that the word 'relatively' is emphasized in the previous sentence. Though the theory underlying the triangular test is much simpler than that underlying the *t*-test, for example, it is not trivial. To fully understand the triangular test and its table of critical values we must explore a new probability distribution. If you do not relish this prospect jump to the next section, but in doing so you may miss a valuable insight into the philosophy of significance testing.

Let us start by *assuming that the null hypothesis is true*. As the null hypothesis states that 'there is no difference in taste between the two toothpastes' it follows that each assessor is seeking a difference that does not exist. Regardless of the thought processes in the minds of the assessors, a selection of the odd-one-out must be a matter of pure chance. Each assessor would therefore have a one third chance of guessing correctly and we could calculate the probability of any number of the twelve assessors picking out the odd one in the three samples placed before him. The probabilities are obtained by substituting suitable values of *n* and *r* into the equation below.

> *Triangular tests*
> The probability of *r* assessors selecting the odd sample in a triangular test involving *n* assessors
>
> $$= \frac{n!}{r!\,(n-r)!} \left(\frac{1}{3}\right)^r \left(\frac{2}{3}\right)^{n-r}$$
>
> if there is actually no difference between the samples.

In Dr Murphy's experiment the probability of eight assessors making a correct selection is equal to:

$$\frac{12!}{8!\,4!} \left(\frac{1}{3}\right)^8 \left(\frac{2}{3}\right)^4 = 0.015$$

With *n* equal to 12 we can let *r* equal any number from 0 to 12 inclusive. Taking these values of *r* one at a time we can calculate the probabilities in Table 6.1.

Table 6.1   Possible results of the experiment

Number of assessors who select the odd sample	Probability
0	0.008
1	0.046
2	0.127
3	0.212
4	0.238
5	0.191
6	0.111
7	0.048
8	0.015
9	0.003
10	0.001
11	0.000
12	0.000

Don't forget that the probabilities in Table 6.1 are based on the assumption that the null hypothesis is true, i.e. that the two types of toothpaste are indistinguishable. The probability distribution that is given in Table 6.1 is also illustrated in Fig. 6.1.

Marked on Fig. 6.1 is the critical value (7.5) for the 5% significance level and we can see that the probability of rejecting the null hypothesis is less than 5%. When Dr Murphy found that eight assessors had identified the odd sample he rejected the null hypothesis at the 5% significance level. The chance of his having made a wrong decision is less than 5%. Had he found nine assessors to

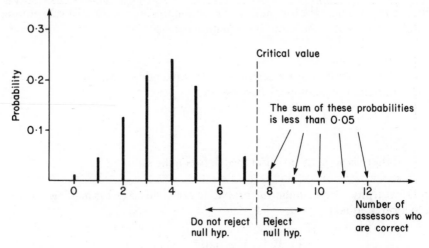

Figure 6.1   What is likely to happen if the toothpastes are indistinguishable

be correct he would have rejected the null hypothesis with even greater confidence, knowing that he incurred a risk which was less than 1%.

The whole table of critical values for the triangular test is based on calculations similar to those which gave the probabilities in Table 6.1. This point might be clearer if we sum the probabilities in table 6.1, starting at the bottom, to obtain the cumulative probabilities in Table 6.2.

The ½ values in the left hand column are introduced in order to avoid complex statements involving phrases such as 'less than or equal to'. Obviously the test statistic must be a whole number in a triangular test so we could have used 3.4 or 3.7 say, rather than 3½ in Table 6.2. The ½ values will enable us to match Table 6.2 with Table D.

To obtain critical values from Table 6.2 we scan the probabilities from top to bottom. The first probability which is less than 0.05 (i.e. 5%) is the 0.019 which is next to 7½. The critical value at the 5% significance level is therefore 7½. Scanning further downwards we find the first probability which is less than 0.01 (i.e. 1%). This is the 0.004 which is next to 8½. The critical value at the 1% significance level is therefore 8½.

Strictly speaking we *could* say that 7½ was the critical value at the 1.9% significance level and that 8½ was the critical value at the 0.4% significance level. It might, however, cause considerable confusion to depart from the respected convention of using 5% and 1% significance levels (with occasional use of 10% and/or 0.1% to extend the repertoire).

Table 6.2   Cumulative probabilities from Table 6.1

Critical value	Probability of more than the above number selecting the odd sample
½	0.992
1½	0.946
2½	0.819
3½	0.607
4½	0.369
5½	0.178
6½	0.067
7½	0.019
8½	0.004
9½	0.001
10½	0.000
11½	0.000

We have examined the theoretical basis of one statistical table. The table of critical values for the triangular test (Table D) is the simplest example we could have chosen. The theory underlying the *t*-table, for example, is considerably more complex and the calculation of the critical values very much more difficult. Fortunately it is possible to use significance tests without any knowledge of how tables of critical values are constructed. Much more important is to be able to select the *appropriate* test and to be aware of the *assumptions* on which it is based. The assumptions underlying some common significance tests will be discussed in the next chapter.

6.4   What percentage of users will detect the additive?

Dr Murphy concluded that it was possible for a person to detect the presence of the new additive PX 235. In his experiment he used trained assessors who were regular members of a tasting panel. Perhaps Dr Murphy should now extend his investigation to include members of the public and whilst doing so he could consider more sophisticated questions than that which he has just answered.

You will recall that the first experiment was a response to the question 'Does the inclusion of PX 235 affect the taste of the toothpaste?' Now that this question has been answered in the affirmative, Dr Murphy realizes that the *nature* of the effect requires to be investigated. It is possible that:

(a) Some people can detect the presence of the additive but others cannot.

(b) The percentage of people who can detect the additive may depend upon its concentration within the toothpaste.

In the first experiment 1 g of PX 235 was added to each kilogram of toothpaste. This is the concentration suggested by the research manager as being necessary for the protection of the tooth enamel. He further maintains that with this level of concentration about 30% of customers would be able to detect the difference between normal and treated toothpastes.

Dr Murphy decides to check this prediction by involving 50 employees in a further experiment. These employees are selected from volunteers who are not normally associated with taste panels and the group includes males and females spread over a wide age range. Each of the 50 is subjected to 24 sets of three samples and asked to select the one which differs from the other two in each set. Only those employees who make 16 or more correct selections will be classified as 'able to detect the difference'. (Note how this experiment differs from that carried out earlier by Dr Murphy. It would be of little use to present each subject with *only one* set of three samples in this second experiment.)

The experiment is carried out and the results show that 23 of the 50 employees (i.e. 46%) *can* detect a difference between the treated and untreated toothpastes. The research manager, when confronted with this evidence, refuses to accept that his suggestion has been disproved. He still maintains that the percentage of customers who could detect the difference may well be 30% and that the 46% found by Dr Murphy is excessively large simply because the employees he has used are not representative of *all tooth-paste users.*

Dr Murphy is constrained by a budget, of course. The cost of selecting a random sample from all toothpaste users and bringing them into the laboratory would be out of all proportion to the benefits gained. But even if he *did* take a random sample there is no guarantee that it would be representative and even if the sample *was* representative it would not be possible for Dr Murphy to demonstrate conclusively that this was so.

It is impossible then for either Dr Murphy or the research manager to prove that the other is wrong. It *is* possible, however, to carry out a significance test using a null hypothesis which is synonymous with the research manager's claim. This test, like all the other significance tests in this book, is based on the assumption that the sample is selected at random from the population referred to in the null hypothesis. It is known as the 'one-sample proportion test' and proceeds as follows:

Null hypothesis – The proportion of toothpaste users who could detect a difference between the treated and untreated toothpaste is equal to 0.3 (i.e. $\pi = 0.3$).

Alternative hypothesis – The proportion of toothpaste users who could detect a difference between the treated and untreated toothpaste is not equal to 0.3 (i.e. $\pi \neq 0.3$).

Test statistic $= \dfrac{p - \pi}{\sqrt{\left[\dfrac{\pi(1-\pi)}{n}\right]}}$

$= \dfrac{0.46 - 0.3}{\sqrt{\left[\dfrac{0.3(0.7)}{50}\right]}}$

$= 2.47$

where p is the sample proportion, π is the population proportion and n is the sample size.

Critical values – From the t-table with infinite degrees of freedom for a two-sided test:

1.96 at the 5% significance level
2.58 at the 1% significance level.

Decision – We reject the null hypothesis at the 5% level of significance.

Conclusion – We conclude that the proportion of toothpaste users who could detect a difference is *not* equal to 0.3.

Note 1: The one-sample proportion test should only be carried out when a large sample has been taken (e.g. $n > 30$).

Note 2: Infinite degrees of freedom are always used in this test.

The significance test is telling us that the results of the experiment offer a strong indication that the research manager is wrong when he claims that only 30% of users would be able to detect the difference between the treated and untreated toothpastes. If, however, we feel that the sample is not representative of the population (as the research manager has suggested) then it would be unwise to adopt this conclusion. It is often said that statistics can be used to 'prove' anything. Perhaps this is a gross overstatement, but it is certainly possible to draw ridiculous conclusions by using a statistical technique without checking the assumptions on which it is based.

If we accept that the research manager is wrong then it would be reasonable to use the data from the experiment to *estimate* the population percentage which we now suspect is *not* equal to 30%. The sample of 50 employees contained 23 (i.e. 46%) who could detect a difference. These figures can be used to calculate a confidence interval for the percentage of all toothpaste users who could detect a difference. When estimating a population proportion by means of a sample proportion we use:

> An approximate confidence interval for a population proportion (π) is given by:
>
> $$p \pm t \sqrt{\left[\dfrac{p(1-p)}{n}\right]}$$
>
> where t is taken from the two-sided t-table with infinite degrees of freedom.

A 95% confidence interval for the proportion of all toothpaste users who could detect a difference is approximately:

$$0.46 \pm 1.96 \sqrt{\left[\frac{0.46(0.54)}{50} \right]} = 0.46 \pm 0.14$$

$$= 0.32 \text{ to } 0.60$$

We can be 95% confident that the percentage of users who could detect a difference lies between 32% and 60%. This is a rather wide interval, and you may be surprised that such a large experiment has produced such an imprecise conclusion. It is unfortunately true that very large samples are needed when we are dealing with qualitative information. Each of the 50 subjects was confronted with 24 sets of samples and the end result is just one 'bit' of information (i.e. whether or not that subject can detect a difference). To measure the person's height, for example, would produce much more information for much less effort, but it would be irrelevant to the purpose of the experiment, of course.

If we want a narrower confidence interval we must take a larger sample. We can, in fact, calculate the size of sample that will be needed to estimate a population proportion to within a certain tolerance ($\pm c$) or the size of sample needed to detect a change ($\pm c$) in a population proportion with a specified level of confidence.

> The size n of sample needed to detect a change c in a population proportion, by means of a one-sample proportion test, is given approximately by:
>
> $$n = p(1-p)(2t/c)^2$$
>
> where t is taken from the two-sided t-table with infinite degrees of freedom and p is an initial estimate of the population proportion. If no initial estimate of the population proportion is available then a conservative estimate of the necessary sample size can be calculated from:
>
> $$n = (t/c)^2$$

Both of the above formulae can be used to calculate the size of sample needed to estimate, by means of a confidence interval, a population proportion to within $\pm c$ if we *divide the resulting sample size (n) by 4.*

If we wish to know the size of sample that we must take in order to be 95% confident of estimating the percentage of users to within $\pm 5\%$ we can substitute $t = 1.96$ and $c = 0.05$ into one of the two formulae. As we already have an estimate of the percentage (46%) we will use the first formula and let

$p = 0.46$ to get:

$$n = p(1-p) (2t/c)^2/4$$
$$= (0.46) (1-0.46) [2(1.96)/0.05]^2/4$$
$$= 381.7$$

Thus a sample size of 382 or more is needed for this purpose. If we wish to be more confident, say 99%, or if we wish to narrow the interval then an even larger sample will be needed. We would, for example, need a random sample of at least 9543 people to be 95% confident of estimating the population percentage to within ±1%. If we did not have the initial estimate (0.46) then the use of the second formula would have been necessary. This would have given 385 and 9604 as the sample sizes.

6.5   Is the additive equally detectable in all flavours?

Dr Murphy has concluded that the presence of the PX 235 additive can be detected by toothpaste users. He has further concluded that approximately 46% of users would be able to detect its presence at a concentration of 1 g/kg, under 'laboratory' conditions. He would like to extend his investigation despite his limited budget. He is particularly anxious to increase his sample size and to include members of the public. The latter innovation would help to satisfy the research manager's obsession with representative samples.

Other researchers in Duostayer are investigating the possibility of introducing different flavours of toothpaste. It occurs to Dr Murphy that his own research could be extended to overlap with that of the 'flavour people', who appear to have more freedom than his own budget will permit. He proposes that his most recent experiment should be repeated using three of the new flavours.

In due course three new flavours, A3, B7 and C2, are selected, then batches of treated and untreated toothpaste are prepared. The actual experiments are carried out as part of a wider market research investigation in six towns in southern England. Dr Murphy is able to attend only two of these sessions but he learns a great deal about life outside the laboratory and he now has more respect for the wisdom of the research manager.

It was intended that 50 subjects would be allocated to each flavour and that

Table 6.3   Results of field experiment on PX 235

Flavour	A3	B7	C2	Total
Number of subjects who could detect the difference	18	13	11	42
Number of subjects who could not detect the difference	22	37	19	78
Total	40	50	30	120

each subject would be presented with 24 sets of samples. Unfortunately the field experiments did not proceed as smoothly as the experiment previously carried out in the laboratory and some results had to be discarded. A summary of the remaining results is contained in Table 6.3.

The proportion of toothpaste users who could detect the presence of the additive varies from flavour to flavour within the sample of 120 users. The proportions for the three new flavours are given in Table 6.4.

Table 6.4   Field experiments

Flavour	A3	B7	C2	All three new flavours
Proportion of users who could detect a difference	0.450	0.260	0.367	0.350

We see from Table 6.4 that in the field experiments only 35% of users could detect a difference between the treated and untreated toothpastes. This figure compares with the 46% recorded in the laboratory experiment. Perhaps the research manager was right when he claimed that approximately 30% of users would detect the presence of the additive.

It would be possible to carry out a 'two-sample proportion test' to compare the proportion in the field experiment (0.35) with the proportion in the laboratory experiment. If we did so we would find a significant difference between the two proportions. This could be taken as support for the research manager's assertion that the sample of 50 employees in the laboratory experiment was not representative of all toothpaste users. Many other interpretations are possible, of course. Perhaps, for example, it really is easier to detect the additive in normal flavour toothpaste than in the experimental flavours.

Because of the difficulty of comparing results from the laboratory with results from the field we will set aside the data from the normal flavour experiment and concentrate on a comparison of the three experimental flavours. To compare three or more sample proportions we use the '*chi-squared test*'.

The chi-squared test is based on frequencies rather than proportions and we already have the ingredients for our test in Table 6.3. To be more precise the numbers in Table 6.3 are known as *observed frequencies*. The basis of the test is to compare the observed frequencies with a set of *expected frequencies*, the latter being calculated on the assumption that the null hypothesis is true. If we find large differences between the observed and expected frequencies we reject the null hypothesis. Let us carry out the test and discuss the finer points later.

Null hypothesis – The proportion of users who can detect the presence of the additive is the *same* for all three flavours.

Alternative hypothesis – The proportion of users who can detect the presence of the additive is *not* the same for all three flavours.

Test statistic – Before we can calculate the test statistic we must first calculate the expected frequencies. To do so we set up a table (Table 6.5) which is very similar to that containing the observed frequencies and has the *same* row totals and column totals.

Table 6.5   For insertion of expected frequencies

Flavour	A3	B7	C2	Total
Number of subjects who could detect the difference				42
Number of subjects who could not detect the difference				78
Total	40	50	30	120

The expected frequencies are inserted in the six empty cells of table 6.5. The following equation is used to calculate the expected frequencies:

$$\text{Expected frequency} = \frac{(\text{row total} \times \text{column total})}{\text{overall total}}$$

Using this equation we get the expected frequencies in Table 6.6.

Table 6.6   Expected frequencies

Flavour	A3	B7	C2	Total
Number of subjects who could detect the difference	14.0	17.5	10.5	42
Number of subjects who could not detect the difference	26.0	32.5	19.5	78
Total	40	50	30	120

Using the expected frequencies in Table 6.6 and the observed frequencies in Table 6.3 we can now calculate the test statistic.

$$\text{Test statistic} = \sum \frac{(\text{observed frequency} - \text{expected frequency})^2}{\text{expected frequency}}$$

$$= \frac{(18 - 14.0)^2}{14.0} + \frac{(13 - 17.5)^2}{17.5} + \frac{(11 - 10.5)^2}{10.5}$$

$$+ \frac{(22 - 26.0)^2}{26.0} + \frac{(37 - 32.5)^2}{32.5} + \frac{(19 - 19.5)^2}{19.5}$$

$$= 3.58$$

Critical values – Taken from the chi-squared table (Statistical Table E) using 2 degrees of freedom:

5.991 at the 5% significance level
9.210 at the 1% significance level.

Note:
Degrees of freedom = (number of rows − 1) (number of columns − 1)

$$= (2-1)(3-1)$$

$$= 2$$

Decision – As the test statistic is less than both critical values we cannot reject the null hypothesis.

Conclusion – We are unable to conclude that the proportion of toothpaste users who can detect the presence of the additive differs among the three flavours.

It is quite possible, then, that the flavour of the toothpaste does not affect a person's ability to detect the presence of the additive. As we have failed to show that the three population proportions are not equal, it is reasonable to accept that they *are* equal and to combine the data from all three flavours to estimate the proportion of users who can detect the presence of the additive. Of the 120 people in the field experiment 42 were able to detect the difference, which gives a sample proportion equal to 0.350 and a 95% confidence interval of 0.350 ± 0.085 for the population proportion.

In carrying out this chi-squared test we used 2 degrees of freedom. In any chi-squared test of this type (there is actually a second type of chi-squared test known as the goodness of fit test) we can calculate the number of degrees of freedom from the equation:

Degrees of freedom = (number of rows − 1) (number of columns − 1)

Alternatively we can obtain the degrees of freedom by counting the minimum number of expected frequencies that *must* be calculated. The expected frequencies are calculated using:

Expected frequency = (row total × column total)/(overall total)

Consider the transition from Table 6.5 to Table 6.6. After we have calculated the 14.0 in the top left cell and then the 17.5 in the adjacent cell we can obtain all the other expected frequencies by subtracting from the row and column totals, which were fixed in advance. Of the six expected frequencies *only two* need to be calculated from the equation, so we have 2 degrees of freedom.

All of the tests in this chapter have been concerned with frequencies or proportions. The data used in each test resulted from *counting* rather than measuring and the variables in question could be described as qualitative rather than quantitative. If you have little or no experience with qualitative variables you may have been surprised to learn that such very large sample sizes are needed when investigating such variables. You may also have been surprised by the difficulties inherent in any situation which involves the assessment of human performance or attitudes. In fairness it should be stated that some of the questions asked by Dr Murphy were not the best questions that he might have asked, nor did he always make best use of the data he had collected. Nonetheless, many important points have been illustrated by the analysis of Dr Murphy's data.

One area in which the industrial scientist or technologist is likely to encounter qualitative variables is in quality control and/or quality assurance. We will close this chapter with an examination of some of the problems which arise in quality assurance.

6.6　The proportion of defective items in a batch

Duostayer Ltd also manufacture baby foods which are packed into glass jars with screw-on metal tops. Different companies supply the jars and the tops, with the jars meeting the tops for the first time in the automatic bottling process. The failure of a jar to mate with its top causes a disruption and a succession of such failures can be very expensive.

Almost all failures result from the diameter of the jar neck being outside specification. In order to reduce the frequency of such occurrences all batches of jars are inspected. To be more precise a *sample* of jars is taken from every batch then each jar in the sample is examined using a go/no-go gauge and the number of *defective* jars in the sample is counted. If the *sample* is found to have very few defective jars the *whole batch* is accepted, but if the number of defectives is excessive the batch is set aside and reported to the supplier.

In order to make a decision about the acceptability of a batch the quality assurance inspector needs to know:

(a) How many jars should he include in the sample?
(b) What is the largest number of defective jars that can be tolerated in the sample, if he is to accept the batch?

These questions are discussed thoroughly in texts on statistical quality control; at this point we will pose a much simpler question 'If a batch of 10 000 jars contains 500 which are defective what is the probability that a random sample of ten jars will contain less than two which are defective?'

Earlier we answered very similar questions when we considered Dr Murphy's first experiment. You may recall that we evaluated the expression:

$$\frac{12!}{8! \, 4!} \left(\frac{1}{3}\right)^8 \left(\frac{2}{3}\right)^4$$

in order to calculate the probability that eight out of twelve assessors would select the odd sample out of three. The very same expression also gives us the probability of finding eight defective jars in a sample of twelve jars taken from a

large batch in which one third of the jars are defective. In fact the formula that
we used when discussing triangular tests, i.e.

$$\frac{n!}{r!\,(n-r)!}\left(\frac{1}{3}\right)\left(\frac{2}{3}\right)^{n-r}$$

can be used in a wide variety of situations if we replace the 1/3 by a symbol. We
would then have a formula for calculating what are known as *binomial*
probabilities.

> *The binomial distribution*
> The probability of getting *r* successes in *n* independent
> trials is
>
> $$\frac{n!}{r!\,(n-r)!}(\pi)^r\,(1-\pi)^{n-r}$$
>
> where π is the probability of success at each trial.

A *trial* can have only *two* outcomes, which are labelled 'success' and 'failure'.
The outcome of each trial is not predictable but there is a fixed chance (π) of
success at each trial. Table 6.7 lists several situations in which it is meaningful to
speak of trials and in which it is valid to use the binomial distribution.
　　Let us return to the question in which we have taken a sample of ten jars from
a large batch. The batch contains 10 000 jars of which 500 are defective so that

Table 6.7   Situations in which the binomial distribution may be useful

Trial	*Success*	*Failure*	*Probability of success at each trial*
A triangular test in which there is no difference between the treated and untreated samples	Select the odd sample	Select other sample	1/3
One toss of a fair coin	Head	Tail	1/2
One roll of a fair dice	Score 'six'	Any other score	1/6
A component selected at random from a large batch	Component is serviceable	Component is defective	Depends on the proportion of serviceable components in the batch
Dialling a particular telephone number, between 10.00 and 10.15 a.m.	Line not engaged	Line engaged	Depends on the proportion of time in which line is not engaged at this time of day

the *proportion* of defective jars in the batch is 500/10 000 which is 0.05. Consider the sampling step by step. The probability that the *first* jar will be defective is 0.05. The probability that the *second* jar will be defective depends to a small extent on the first jar. If the first was defective we have a probability of 499/9999 for the second, but if the first jar was not defective we have a probability of 500/9999. Clearly these two probabilities differ (0.0499049 against 0.050005), but the difference is negligible. We will assume therefore that the probability of a jar being defective is 0.05 for *each* of the ten jars in the sample. In the words of the binomial distribution we will assume that the probability of success at each trial is equal to 0.05 (i.e. $\pi = 0.05$).

To find the probability of getting one defective jar in a sample of ten we further let $n = 10$ and $r = 1$, then we can calculate:

Probability of one defective jar in a sample of ten

$$= \frac{10!}{1! \; 9!} (0.05)^1 \, (0.95)^9$$

$$= 0.315$$

To obtain the probability of getting zero defective jars in a sample of ten we use $n = 10$, $r = 0$ and $\pi = 0.05$.

Probability of zero defective jars in a sample of ten

$$= \frac{10!}{0! \; 10!} (0.05)^0 \, (0.95)^{10}$$

$$= 0.599$$

Adding these two results we obtain the probability of getting *less than two* defective jars in a sample of ten:

$$0.315 + 0.599 = 0.914$$

6.7 Summary

In this chapter we have examined some of the significance tests which have proved useful when dealing with *qualitative* variables. The triangular test is used in sensory evaluation whilst the one-sample proportion test and the chi-squared test are widely used in the analysis of survey data as well as large scale experiments. We noted that qualitative data does not contain as much information as quantitative measurements and therefore large samples are needed if we are to detect small differences or make precise estimates.

In one example the conclusions of the experimenter were criticized on the grounds that his sample was not representative of the population being investigated. This is a theme to which we will return in the next chapter when we discuss the assumptions on which significance tests are based.

Problems

(1) In order to test a colour-matcher's ability to discriminate, a triangular test was arranged with two fabrics whose difference in colour was only a 'shade'. The colour matcher picked the 'odd-one-out' correctly seven times out of fifteen.

(a) Can the colour-matcher discriminate between the fabrics?
(b) Is this a suitable design?

(2) With electrical insulators intended for outdoor use, the surface finish of the insulator is extremely important. In the manufacture of a particular high quality insulator the surface coating is formed by heating as the insulators pass through a continuous belt furnace. Every insulator is visually inspected as it leaves the furnace to detect 'hazing' of the surface. We know from experience that even when the furnace is working at its best we can expect 10% of insulators to be hazed. If the operating conditions of the furnace change from the optimum, the percentage of insulators which are hazed will be higher than 10%. Hazing appears to occur at random, and no 'serial patterns' of hazing have ever been substantiated.

(a) Examination of the inspection results from a single day's production revealed that 31 insulators were hazed out of a total of 200. Carry out a significance test to confirm that the furnace has deteriorated below its optimum.
(b) Calculate a 95% confidence interval for the true proportion of hazed insulators.
(c) What is the population for the significance test used in part (a)?
(d) A closer analysis of the results of the day's production revealed that two inspectors were used and that inspector A found only eight hazed insulators out of the 80 which he examined. The remainder of the 200 insulators were examined by inspector B. Complete the following table:

	Inspector		
Insulator	*A*	*B*	*Total*
Hazed			
Not-hazed			
Total			200

(e) Carry out a test using the table in part (d) to check whether the inspectors have obtained a significantly different proportion of hazed insulators.
(f) Use the binomial distribution to complete the following table which refers to the probability of obtaining hazed insulators for a sample of size ten with the furnace operating at its optimum.

Number of hazed insulators	0	1	2	3	4	5 and above
Probability	0.349		0.194	0.057	0.011	

What is the probability of finding two or more hazed insulators in a sample of ten under this condition?

(g) Use the binomial distribution to complete the following table which refers to the probability of obtaining hazed insulators for a sample of size ten when the furnace is producing 20% hazed insulators.

Number of hazed insulators	0	1	2	3	4	5 and above
Probability		0.268	0.302	0.201		0.034

What is the probability of finding less than two hazed insulators in the sample when the furnace is producing 20% hazed insulators?

(h) It is decided to control the quality of the day's production using the first ten insulators. If there are less than two hazed insulators in this sample it is concluded that the furnace is working at its optimum, otherwise it is concluded that the furnace has deteriorated.

(i) What is the probability of concluding that the furnace has deteriorated when it is working at its optimum?

(ii) What is the probability of concluding that the furnace is working at its optimum when the true proportion of hazed insulators is 20%?

What conclusions can be drawn about the above decision criteria?

7

Testing and estimation: assumptions

7.1 Introduction

After reading the last three chapters you may well be convinced that it is easy to carry out a significance test. True, there is the initial difficulty of deciding *which* test to do and then there is the chore of having to think about the population when it is so much easier to confine one's attention to the sample, but these problems shrink into insignificance with practice.

Unfortunately there is another problem which has been carefully avoided in earlier chapters but will now be brought out into the open. This concerns the assumptions on which significance tests are based. It is essential that the user of a significance test should be aware of the assumptions which were used in the production of the table of critical values. Any statistical table is intended to be used in certain circumstances. The use of a table of critical values in a situation for which it was not intended can lead to the drawing of invalid conclusions.

Before we probe into assumptions, however, we will examine more sophisticated ways of estimating the variability which must be taken into account when doing a *t*-test.

7.2 Estimating variability

In Chapter 5 we used the *F*-test to compare the precision of two analytical test methods. Both methods had been used to obtain repeat measures of impurity on the same batch and the standard deviation of the five results by method 1 was 0.3341 whilst the standard deviation of the six results by method 2 was 0.2190. The ratio of the two variances was less than the 5% critical value and we were therefore unable to conclude that the precision of the second method was superior to that of the first. We then went on to explore the possible bias of the two methods using a *t*-test.

In passing so quickly from the *F*-test to the *t*-test we missed the opportunity to *estimate* the variability of each test method. We will now do so, by using the two sample standard deviations to calculate 95% confidence intervals for the population standard deviations. To do this we make use of the following formula:

A confidence interval for a population standard deviation is given by:

Lower limit $= L_1 s$

Upper limit $= L_2 s$

where L_1 and L_2 are obtained from Statistical Table F.

Five results were obtained by method 1 and therefore the sample standard deviation(0.3341) has 4 degrees of freedom. Table F gives us $L_1 = 0.60$ and $L_2 = 2.87$ for a 95% confidence interval with 4 degrees of freedom.

$$\text{Lower limit} = 0.60 \times 0.3341 = 0.200$$

$$\text{Upper limit} = 2.87 \times 0.3341 = 0.959$$

We can therefore be 95% confident that the population standard deviation lies between 0.200 and 0.959. If we were to continue repeatedly measuring the impurity of this particular batch using this test method we could be confident that the standard deviation of the impurity measurements would eventually settle down to a figure between 0.200 and 0.959.

With the second test method six measurements were made, giving a standard deviation of 0.2190 with 5 degrees of freedom. Table F gives us $L_1 = 0.62$ and $L_2 = 2.45$ for a 95% confidence interval with 5 degrees of freedom.

$$\text{Lower limit} = 0.62 \times 0.2190 = 0.136$$

$$\text{Upper limit} = 2.45 \times 0.2190 = 0.537$$

We can, therefore, be 95% confident that the population standard deviation for the second method of test lies between 0.136 and 0.537. The two confidence intervals are depicted in Fig. 7.1 which illustrates two points very clearly:

(a) There is considerable overlap between the two intervals.
(b) Neither of the confidence intervals is symmetrical (i.e. the sample standard deviation is *not* in the centre of the interval).

It should not surprise us that there is overlap between the two confidence intervals since we have already *failed* to demonstrate, by means of the *F*-test, that the two population variances are unequal. The absence of symmetry in the confidence intervals can come as a surprise to many who are familiar with the more common confidence interval for the population mean. The asymmetry arises because the sampling distribution of sample standard deviations is skewed. (For a discussion of sampling distributions see Appendix C.)

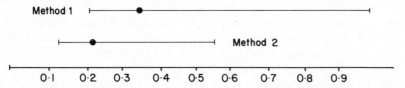

Figure 7.1   Confidence intervals for population standard deviations

A third point which may have struck you forcefully when looking at Fig. 7.1 is the enormous width of the confidence intervals. Had the samples been larger the intervals would have been narrower, of course. A browse down the columns of Table F will soon convince you, however, that a very large sample would be needed to get a much narrower interval. For example a sample standard deviation of 0.3 would give a 95% confidence interval from 0.255 to 0.366 if based on 60 degrees of freedom. At the other extreme the same standard deviation (0.3), if based on only 1 degree of freedom, would give rise to a 95% confidence interval which extended from 0.135 to 9.57.

7.3 Estimating variability for a *t*-test

When carrying out a one-sample *t*-test we calculated the test statistic using:

$$\frac{\bar{x} - \mu}{s/\sqrt{n}}$$

The value of the population mean (μ) was not known, but a value was given to μ in the null hypothesis. The other three ingredients ($\bar{x}$, s and n) were obtained from the sample. We had taken one sample which yielded n observations and the values of $\bar{x}$ and s were calculated from these observations.

It is *not necessary*, however, to calculate the standard deviation (s) and the mean ($\bar{x}$) from the *same* sample. In fact there are many situations in which it is *desirable* to obtain $\bar{x}$ and s from *different* sets of data. Perhaps the most common example occurs when a sample of data has been obtained in order to explore a possible change in mean (μ). Clearly it is possible to calculate $\bar{x}$ and s from these data then carry out a one-sample *t*-test. Whilst this would be quite valid it may not be the *best* course of action, especially if the sample is small. If we are able to *assume* that a change in mean (μ) will *not* have been accompanied by a change in standard deviation (σ) *and* we have an estimate of σ from past data then it may be better to use this estimate rather than the sample standard deviation (s). It will certainly be tempting to use this estimate if it is based on a *much larger number of degrees of freedom* than the ($n - 1$) on which the sample standard deviation is based. We will be in a better position to detect a change in mean (μ) if we have an estimate of σ based on a *large* number of degrees of freedom than we would be if we were using the standard deviation of a *small* sample.

Estimates of population standard deviations do not occur by magic! They can only come from data gathered in the past. We are *not* advocating that you should make indiscriminate use of *all* available data in a mad scramble to boost your degrees of freedom. We are simply advising that you should make the *best* use of any available data which are known to be suitable.

In some situations it is possible to group together the data from *several* small samples in order to obtain a good estimate of the population standard deviation. The same practice·can be followed if we have one sample from each of several populations which are known to have the same standard deviation. The point will be illustrated by the following example.

Example 7.1 The research centre of a well known oil refining company has been asked to investigate the biodegradation of hydrocarbons in water. The aim of the investigation is to compare the performance of several micro-organisms and to examine ways of accelerating the degradation process. Each experiment involves the introduction of a specified level of a hydrocarbon into a vessel of water and the measurement of the level at regular intervals over a three-month period. The determination of the hydrocarbon level in a sample of water is a time consuming and expensive operation, with each determination requiring the undivided attention of a laboratory assistant for approximately three hours. It is only practicable therefore to perform two, or at the most three, repeat determinations on each sample.

In an experiment designed to compare six micro-organisms, a concentration of 25 parts per million of hydrocarbon is introduced into each of six vessels. Fourteen days later the hydrocarbon levels are found to be as in Table 7.1.

Table 7.1 Hydrocarbon level after 14 days

Vessel	A	B	C	D	E	F
Determination of	12.2	8.6	17.4	13.0	14.6	12.7
hydrocarbon level	14.3	10.4	16.1	16.2	11.3	12.9
(p.p.m.)		9.9			13.1	
Mean	13.25	9.63	16.75	14.60	13.00	12.80
Standard deviation	1.485	0.929	0.919	2.263	1.652	0.141

As a first step in the analysis of the data in Table 7.1 it is required to calculate a 95% confidence interval for the mean hydrocarbon level in each vessel.

The calculation of a confidence interval for a population mean is not difficult. We first attempted the task in Chapter 4, when we made use of the formula:

$$\text{Confidence interval for } \mu \text{ is } \quad \bar{x} \pm ts/\sqrt{n}$$

Using this formula with the data in Table 7.1 we will need $t = 12.71$ for vessels A, C, D and F, on which only two determinations were made, and $t = 4.30$ for vessels B and E, on which three determinations were carried out. Each vessel is treated quite separately from the other five vessels and the 95% confidence intervals are as in Table 7.2.

Several of the confidence intervals in Table 7.2 are staggeringly wide. This is a consequence of basing a confidence interval on such little information. With vessels A, C, D and F we have used the *absolute minimum* number of observations ($n = 2$) from which a confidence interval can be calculated and the resulting t value of 12.71 for 1 degree of freedom gives rise to the very wide intervals. (The interval for vessel F is surprisingly narrow, however, because of

Table 7.2   95% confidence intervals for true hydrocarbon level

Vessel	Mean	SD	t	Confidence interval	
A	13.25	1.485	12.71	13.25 ± 13.35	−0.10 to 26.60
B	9.63	0.929	4.30	9.63 ± 2.31	7.32 to 11.94
C	16.75	0.919	12.71	16.75 ± 8.26	8.49 to 25.01
D	14.60	2.263	12.71	14.60 ± 20.34	−5.74 to 34.94
E	13.00	1.652	4.30	13.00 ± 4.10	8.90 to 17.10
F	12.80	0.141	12.71	12.80 ± 1.27	11.53 to 14.07

the very small sample standard deviation.) Fig. 7.2 illustrates the six confidence intervals and the determinations on which they are based.

Visual inspection of Fig. 7.2 might give you the impression that the confidence intervals are unduly pessimistic. Common sense may suggest to you that the calculated intervals are *not* a fair reflection of the uncertainty which prevails. Surely the cause of our uncertainty results from the fact that repeat determinations on any vessel do not give identical results. Whether this is

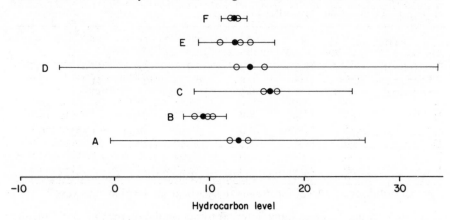

Figure 7.2   Confidence intervals for true hydrocarbon level

caused by sampling variation or by testing error or by some other factor it is *the same cause which is affecting all six vessels*. It is reasonable therefore to calculate *one* estimate of variability and to use it when calculating each of the six confidence intervals. The standard deviations of the six samples can be combined using the following formula:

Combined estimate of the population standard deviation

$$s_c = \sqrt{\frac{\Sigma(\text{d.f.} \times s^2)}{\Sigma \text{ d.f.}}}$$

where d.f. is degrees of freedom and
s is sample standard deviation.

The combined estimate of σ is given by

$$\sqrt{\{[(1 \times 1.485^2) + (2 \times 0.929^2) + (1 \times 0.919^2)}$$
$$+ (1 \times 2.263^2) + (2 \times 1.652^2) + (1 \times 0.141^2)]/$$
$$[1 + 2 + 1 + 1 + 2 + 1]\}$$
$$= \sqrt{\frac{15.36}{8}}$$
$$= 1.386$$

This combined estimate of σ is simply the square root of a weighted average of the sample variances. Not surprisingly then its value lies between the smallest sample standard deviation (0.141) and the largest (2.263). The important feature of this combined estimate of σ is the fact that it has *8 degrees of freedom*. Combining the sample standard deviations in this way has given us an estimate which is as good as one would get from *nine* repeat determinations on any one vessel. When using this estimate to calculate a confidence interval we would therefore use $t = 2.31$ which is the appropriate value for 8 degrees of freedom.

Table 7.3   95% confidence intervals using the combined estimate of σ

Vessel	Mean	SD	t	Confidence interval	
A	13.25	1.386	2.31	13.25 ± 2.26	10.99 to 15.51
B	9.63	1.386	2.31	9.63 ± 1.85	7.78 to 11.48
C	16.75	1.386	2.31	16.75 ± 2.26	14.49 to 19.01
D	14.60	1.386	2.31	14.60 ± 2.26	12.34 to 16.86
E	13.00	1.386	2.31	13.00 ± 1.85	11.15 to 14.85
F	12.80	1.386	2.31	12.80 ± 2.26	10.54 to 15.06

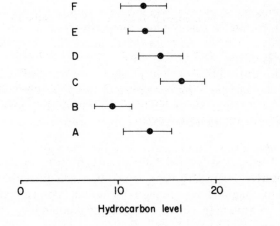

Figure 7.3   Revised confidence intervals for true hydrocarbon level

Recalculating the 95% confidence intervals we obtain Table 7.3. The confidence intervals in this table are illustrated in Fig. 7.3. Comparing these intervals with those depicted in Fig. 7.2 we see considerable differences. Using the combined estimate of the population standard deviation has given us much narrower intervals which are surely a better indication of the uncertainty in our estimates of the six population means.

This example has taken us one step beyond the straightforward application of the one-sample *t*-test, towards a very powerful technique known as 'analysis of variance'. This is discussed in more detail in Part Two of this book.

7.4   Assumptions underlying significance tests – randomness

In *all* of the significance tests described in this book, it is assumed that the sample from which the test statistic is calculated is *drawn at random* from the population referred to in the null hypothesis. To obtain a *random sample* we must engage in *random sampling* which was earlier described as any process which ensures that *every* member of the population has the *same* chance of being included in the sample. To complete the definition we must add that every possible sample of size *n* must have the same chance of being selected.

One way to ensure these *equal chances* is to number each member of the population and then to select by means of *random numbers* taken from tables which are readily available.

If the population is infinitely large it will not be possible to number each member and it is difficult to imagine how one might take a random sample from such a population. In other situations it will not be possible to obtain a random sample because some members of the population do not exist at this point in time.

It may be possible to circumvent some of these difficulties by *redefining* the population in such a way that it becomes possible to take a random sample from the redefined population. Unfortunately this can be self-defeating as the conclusions may become so particular as to be of no interest whatsoever. Perhaps it is more useful to have a non-random sample from an interesting population than to have a random sample from a population which has been so narrowly defined that it is sterile.

It seems intuitively obvious that a significance test will lead us to a valid conclusion if the sample is *representative* of the population. Perhaps significance testing would be easier to understand if statisticians spoke of representative samples rather than random samples. Unfortunately there are two major obstacles:

(a) It does not seem possible to define satisfactorily what we mean by the word 'representative', especially when speaking of a small sample.
(b) It is hard to imagine how one should set about taking a sample in order to get representativeness, without resorting to randomness.

Despite the difficulty, or even impossibility, of obtaining a random sample in many situations we are stuck with the assumption underlying all significance tests that a random sample has been taken from the population in question. If you cannot take a random sample what are you to do? Perhaps the best course of action is to carry out the most appropriate significance test and then to qualify your conclusions with comments on the possible representativeness of the sample.

Before we pass on to other assumptions it must be stated that in many areas of research a random sample is the exception rather than the rule. Charles Darwin and Isaac Newton managed to draw conclusions of great generality without the benefit of random sampling. Darwin's theory of evolution relates to *all* living creatures but he never attempted to take a random sample from even one species. Newton didn't need to be bombarded by a random sample of apples before he saw the light.

7.5 Assumptions underlying significance tests – the normal distribution

The *F*-test and the various types of *t*-test are all based on the assumption that the observations have been taken from a normal distribution. In other words, it is assumed that the population referred to in the null hypothesis has a normal distribution.

When we discussed probability distributions in Chapter 3 it was pointed out that the normal distribution is a mathematical model which, strictly speaking, is only applicable to certain, very abstract situations. On the other hand it is also true that:

(a) many sets of data give histograms which resemble normal distribution curves;
(b) the normal distribution has proved its usefulness for describing variability in an enormous variety of situations.

In practice, then, the *F*-test and the *t*-test are used in circumstances where, strictly speaking, they are not applicable. The user of significance tests should always question the assumptions underlying a test but it is surely better to ask 'How badly are the assumptions violated?' rather than 'Are *all* the assumptions satisfied?'

A further question which the user might ask of a significance test is 'To what *extent* is this test dependent on the assumptions being satisfied?' In doing so he would be questioning what is known as the robustness of the significance test. All statistical techniques are based on assumptions. A *robust* technique is one which is not very dependent upon its assumptions being satisfied. When using a technique which is robust we are likely to draw a valid conclusion even if the assumptions are quite badly violated. On the other hand, when using a technique which is *not* robust we are likely to draw invalid conclusions if the assumptions are not well satisfied.

It is difficult to quantify robustness. The *t*-test, however, is generally regarded as being quite robust as far as the normality assumption is concerned, especially if the sample is large. The *F*-test, unfortunately, is known to be very dependent on the normality assumption. Despite this well-known fact the *F*-test is widely used especially in the computer analysis of large sets of data, as we will see in Part Two of this book. A technique which is less than perfect may nonetheless be very useful especially if used with caution.

The *t*-test, then, is not very dependent on the assumption of an underlying normal distribution. You can feel quite safe in using the *t*-test, provided the population distribution is not very skewed. It is always possible to check, by examination of the sample distribution, whether or not the population distribution is reasonably close to the shape of a normal curve. This can be done by carrying out a 'chi-squared goodness to fit test'. Unfortunately this is a much bigger task than the *t*-test itself and in practice it is much more convenient to examine only the 'tails' of the sample distribution. For this purpose there exist a variety of tests known as 'outlier tests' or 'tests for outliers'. These will be considered later.

7.6   Assumptions underlying significance tests – sample size

In the previous chapter we used the triangular test and the one-sample proportion test. Though the two tests were used in slightly different situations you may have spotted a similarity and it may have occurred to you that the triangular test was only a special case of the one-sample proportion test in which the null hypothesis was $\pi = 1/3$. This is not quite true, however, because of the assumptions concerning the sample size in the two tests. Whilst the one-sample proportion test must only be used with a large sample (i.e. greater than 30) the triangular test can be used with any size of sample. It would only be valid to replace the triangular test with a proportion test, therefore, if the sample were large.

Many statistics books refer to the triangular test as an 'exact test' and to the proportion test as an 'approximate test'. Other books make a slightly different distinction, between 'small sample tests' on the one hand and 'large sample tests' on the other.

Regardless of how the material is presented there are several significance tests which depend upon the fact that certain probability distributions can be approximated by the normal distribution. Unfortunately the approximation will only be a good one if certain conditions are satisfied. The one-sample proportion test is an approximate test based upon the assumption that a binomial distribution can be approximated by a normal distribution. It can be shown that the approximation is better:

(a) with larger samples;
(b) with values of π close to 0.5.

It is generally agreed, therefore, that the one-sample proportion test should only be used if the sample size (n) is at least 30 *and furthermore* both $n\pi$ and

$n(1 - \pi)$ are greater than 5. Thus it would be reasonable to test the null hypothesis $\pi = 0.2$ with a sample size of 30 but to test $\pi = 0.1$ we would need a sample of at least 50.

Another approximate test dealt with in the previous chapter is the chi-squared test. This again is dependent upon a 'normal approximation' and for this reason *a chi-squared test should never be carried out if any of the expected frequencies are less than 5*. Clearly if we are to satisfy this condition our sample size will need to exceed a minimum figure. For example, if we have a two-way frequency table with six cells (like Table 6.5) then we cannot have all six expected frequencies greater than 5 unless our sample size exceeds 30.

7.7 Outliers

The idea of an 'outlier' appeals to the practical scientist/technologist. It seems intuitively obvious to him or her that a set of data may contain one or more observations which need to be singled out as 'not belonging' and may need to be discarded before analysis can begin. We will define an *outlier* then as an observation which does not appear to belong with the other observations in a set of data.

Picking out observations which 'do not appear to belong' is a very subjective process and, when practised by the unscrupulous (or the weak), could lead to undesirable results. Fortunately statisticians have devised significance tests which can be used to determine whether or not an 'apparent outlier' really is beyond the regular pattern exhibited by the other observations. Tests for outliers, like all other significance tests, are carried out using the purely objective, six-step procedure. They can, nonetheless, be abused by persons who:

(a) either make use of outlier tests only when it suits their purpose to do so;
(b) or are unaware of the assumptions underlying outlier tests;
(c) or reject outliers indiscriminately, without first seeking non-statistical explanations for the cause of the outlier(s).

Underlying *every* test for outliers is the usual assumption that the sample was selected at random. In addition there is an assumption concerning the distribution of the population from which the sample was taken. This second assumption is not appreciated by everyone who makes use of outlier tests and ignorance of this assumption can lead to the drawing of ridiculous conclusions.

One of the best known of all outlier tests is 'Dixon's test', the use of which will be illustrated with reference to the data introduced in Chapter 2. To clarify the presentation we will consider only the first six batches of pigment and the yield and impurity of these batches is given in Table 7.4.

You may recall that in Chapter 4 we used the yield values of the first six batches to test the null hypothesis that the mean yield was still equal to 70.3. Many statisticians would advocate that *before* carrying out this *t*-test we should have used Dixon's test to check for outliers.

Table 7.4   Yield and impurity of six batches of pigment

Batch	Yield	Impurity
1	69.0	1.63
2	71.2	5.64
3	74.2	1.03
4	68.1	0.56
5	72.1	1.66
6	64.3	1.90

To calculate the test statistic we first rearrange the observations into ascending order, with the smallest value (x_1) at the left and the largest value (x_n) at the right. For the yields of the six batches this would give:

x_1	x_2	x_3	x_4	x_5	$x_6(x_n)$
64.3	68.1	69.0	71.2	72.1	74.2

The formula used to calculate the test statistic depends upon the sample size, as in Table 7.5.

Table 7.5   Test statistic for Dixon's test

The value of the test statistic is given by A or B, whichever is the greater.

Sample size	A	B
3 to 7	$\dfrac{x_2 - x_1}{x_n - x_1}$	$\dfrac{x_n - x_{n-1}}{x_n - x_1}$
8 to 12	$\dfrac{x_2 - x_1}{x_{n-1} - x_1}$	$\dfrac{x_n - x_{n-1}}{x_n - x_2}$
13 or more	$\dfrac{x_3 - x_1}{x_{n-2} - x_1}$	$\dfrac{x_n - x_{n-2}}{x_n - x_3}$

For the yields of the six batches we calculate:

$$A = \frac{x_2 - x_1}{x_6 - x_1} = \frac{68.1 - 64.3}{74.2 - 64.3} = 0.384$$

$$B = \frac{x_6 - x_5}{x_6 - x_1} = \frac{74.2 - 72.1}{74.2 - 64.3} = 0.212$$

Test statistic = greater of A or B = 0.384

Critical values – Are taken from Statistical Table G and for a sample size of six are:

0.628 at the 5% level of significance
0.740 at the 1% level of significance.

Decision – We cannot reject the null hypothesis.

Conclusion – We are unable to conclude that the yields of the six batches contain an outlier.

The reader will have noticed that the Dixon's test was carried out without a null hypothesis or an alternative hypothesis. Obviously this is unwise and we will return to this point later.

We can, of course, apply the same procedure to the impurity determinations of the six batches. If we intended to use a one-sample *t*-test to make a decision about the mean impurity of all batches, it would be reasonable to precede the *t*-test with a Dixon's test. Putting the impurity determinations into ascending order gives:

x_1	x_2	x_3	x_4	x_5	x_6
0.56	1.03	1.63	1.66	1.90	5.64

$$A = \frac{x_2 - x_1}{x_6 - x_1} = \frac{1.03 - 0.56}{5.64 - 0.56} = 0.093$$

$$B = \frac{x_6 - x_5}{x_6 - x_1} = \frac{5.64 - 1.90}{5.64 - 0.56} = 0.736$$

Test statistic = greater of A or B = 0.736

Critical values – are taken from Table G and for a sample size of six are:

0.628 at the 5% significance level
0.740 at the 1% significance level.

Decision – We reject the null hypothesis at the 5% level of significance.

Conclusion – We conclude that the impurity determinations of the six batches do contain an outlier.

A graphical representation of the data in Fig. 7.4 clearly indicates which of the six determinations does not appear to belong with the other five.

When we say that the 5.64 does not appear to belong with the other five determinations we are basing this judgement on our preconception of how we *expect* data to cluster together. It may be difficult, or even impossible, to state explicitly what we *do* expect but the extreme value (5.64) only appears to be an outlier because it *conflicts* with what we expect to find when we make six observations. Dixon's test is also based on a preconception, or assumption. With a significance test it is possible, indeed essential, to state explicitly the

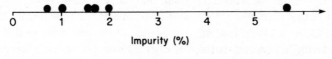

Impurity (%)

Figure 7.4

assumptions on which it is based. *Dixon's test is based on the assumption that the observations have come from a normal distribution.* (There are actually other versions of Dixon's test based on other distributions.)

Perhaps at this point it would be wise to correct the earlier omission, by stating a null hypothesis and an alternative hypothesis for the Dixon's test. These hypotheses can be presented in such a way that the assumption of normality is clear:

Null hypothesis – All six yield values came from the same population, which has a normal distribution.

Alternative hypothesis – Five yield values came from the same population, which has a normal distribution, whereas the most extreme yield value does not belong to this population.

Changing the word 'yield' to 'impurity' would give us the hypothesis for the second Dixon's test.

Having concluded that the impurity value of 5.64 is an outlier it would not be valid to carry out a *t*-test on these data, and we now have *two* difficulties to overcome:

(a) How are we to reach a decision about the mean impurity of all batches, now that we know the *t*-test is invalid?
(b) What do we do with the extreme impurity determination (5.64)?

One solution to both problems would be to ignore the outlier and to press on with the *t*-test. This is at best unwise and at worst dishonest.

An alternative solution to both problems would be to *reject* the outlier and to carry out a *t*-test using the other five determinations. Perhaps this is the most natural reaction to an outlier – reject it! In this case, however, it would be a ridiculous course of action.

This assertion is made in the light of further information. We have already examined the impurity determinations of the first 50 batches in Chapter 2. The distribution of impurity is illustrated in Fig. 2.2, which depicts a very *skewed* distribution. The use of Dixon's test has in effect proved what we already know, that the variation of impurity from batch to batch does *not* follow a normal distribution. You may recall that in Chapter 2 we transformed the impurity data by taking logarithms and that the result of the transformation was to give a symmetrical distribution.

We have therefore a third course of action that we might take. Transformation of the impurity data would appear to be more sensible than either ignoring or rejecting the outlier. If we were to carry out a Dixon's test on the six log impurity values we would conclude that there were no outliers in the data and it would then be valid to carry out a *t*-test using the transformed data.

In conclusion we note that *all* significance tests mentioned in this book are based on assumptions concerning the distribution of the population referred to in the null hypothesis. In many cases these assumptions can be checked by means of a test for outliers. We can, for example precede a one-sample *t*-test with a Dixon's test. It is important to be aware, however, that the outlier test itself is based on an assumption about the population distribution. In fact there exists an enormous variety of outlier tests which are discussed at length in Barnett and Lewis (1979).

7.8 Summary

In this chapter we have discussed the assumptions which underlie significance tests *and* confidence intervals. These assumptions fall into two sets, which relate to:

(a) how the sample is selected from the population;
(b) the distribution of the population from which the sample is taken.

All of the tests and confidence interval formulae covered in this book are based on the assumption that a random sample has been taken and most of the tests share the further assumption that the population has a normal distribution. These assumptions are very restricting but they cannot be ignored. Many researchers, especially in the social sciences, find that these assumptions cannot be satisfied and turn to a set of significance tests known as 'non-parametric tests'. These are dealt with in a later volume, *Further Significance Testing*.

Problems

(1) A random sample of size eight was chosen from a week's production of a particular make of car. Road tests on the cars gave the following m.p.g.:

30.7　32.4　32.3　34.1　27.3　32.5　31.7　32.7

(a) Carry out Dixon's test to determine whether the data contain an outlier.
(b) Should the outlier be rejected?

(2) **Source:** *The Analyst*, Jan. 1980, p. 71.

In a collaborative study, each of five laboratories analysed the same grass sample (2 g) for antimony, using the nitric-acid/sulphuric-acid digestion method. Each laboratory carried out three replicate determinations to obtain the results in the table on the following page.

(a) Calculate a combined estimate of the standard deviation.
(b) Use this estimate to obtain a 95% confidence interval for laboratory B.
(c) For each laboratory use Dixon's test to determine whether there is a possible outlier.
(d) By examining the pattern of results given in Fig. 7.5, consider whether the analyses in parts (a), (b) and (c) fairly represent the data.

Laboratory	Determination (µg antimony)			Mean	SD
A	2.34	2.10	1.96	2.13	0.192
B	1.97	1.92	1.91	1.93	0.032
C	2.01	2.01	2.07	2.03	0.035
D	2.06	2.02	2.12	2.06	0.050
E	2.38	1.77	2.47	2.20	0.380
F	1.94	2.06	2.06	2.02	0.069

(e) Which of the following assumptions may have been violated in parts (a), (b) or (c):

  (i) The results follow a normal distribution.
  (ii) The population standard deviations are equal.
  (iii) The results are independent of each other.
  (iv) A random sample has been taken.
  (v) A continuous variable has been measured.

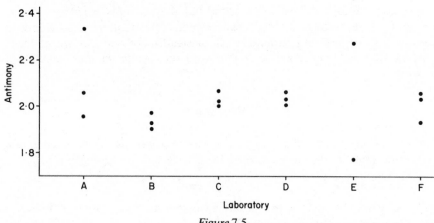

Figure 7.5

(3) On 11 March 1980, *Newsnight*, a BBC television programme, carried out an investigation into the death rate due to cancer at the Atomic Weapons Research Establishment at Harwell. One worker who had recently died of cancer had been subjected to whole-body monitoring for plutonium radiation level at four different time periods. The following radiation levels were obtained:

Time period	1	2	3	4
Radiation level	58	16	0	52

A level of above 50 is considered dangerous and a level of 16 is considered acceptable. After the first and fourth time periods the worker was removed from contact with radiation.

(a) Calculate a 95% confidence interval for the mean radiation level.
(b) Calculate a 95% confidence interval for the true proportion of results above 50.
(c) Which of the following assumptions (or conditions) have been violated in parts (a) or (b):

   (i)   results are a random (or representative) sample of the population;
   (ii) results are independent of each other;
   (iii) the results follow a normal distribution;
   (iv) the number of 'successes' is greater than five. This is a necessary condition for the use of the formula:

$$p \pm t_\infty \sqrt{[p(1-p)/n]}$$

(d) In the light of your conclusions from (c) comment on the analysis in parts (a) and (b).

Part Two

─────8─────
Investigations and statistics

8.1 Introduction

Much research and development work is concerned with *investigating* production processes and other complex systems. Frequently such investigations involve sampling and often take place in an environment in which there is random variation. It is not surprising then, that the scientist or technologist finds use for statistics when he is attempting to reach decisions about cause–effect relationships within the system under investigation.

Though the techniques of statistical analysis may not be used until the later stages of an investigation it is highly desirable that many of the ideas put forward in this book should be taken into account during the planning stage, before any data are obtained. Perhaps this point will be made clear if we consider the methodological framework in which an investigation takes place. Fig. 8.1 illustrates what is often referred to as the scientific method or the hypothetico-deductive method.

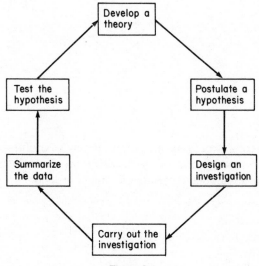

Figure 8.1

Not all scientists would accept that Fig. 8.1 is a valid model of the scientific approach but it is probably true to say that this model could command as much acceptance as any other. Certainly Fig. 8.1 is just as acceptable to a *survey* as it is to an *experiment* and just as applicable to an investigation carried out in a *laboratory* as it is to one carried out in the *'field'*. In this book we will be concerned with experiments in which variables are deliberately changed by the experimenter rather than with surveys in which changes in the variables occur for other reasons. We will pay particular attention to three of the boxes in Fig. 8.1:

(a) Design an investigation.
(b) Summarize the data.
(c) Test the hypothesis.

For data summary and hypothesis testing we will make use of a very powerful statistical technique known as multiple regression analysis. An attempt will be made to illustrate the dangers inherent in the indiscriminate use of this technique and ways of avoiding certain pitfalls will be demonstrated. On the other side of the coin we will discuss some basic principles of experimental design. We will attempt to analyse the results from several well-designed experiments and the results from two badly-designed experiments. Our inability to draw clear cut conclusions from the latter will serve to emphasize the need for statistical wisdom at the planning stage of an investigation.

8.2   Some definitions

The performance of a production process or the outcome of a laboratory experiment can often be quantified in terms of one or more *responses* or *dependent variables*. These responses might include:

(a) yield
(b) taste
(c) throughout
(d) impurity
(e) brightness
(f)   shelf life
(g) cycle time
(h) quality.

In order to control a process or a reaction, it may be useful to understand the effect on a response of one or more *independent variables* (or factors or treatments). These independent variables might include:

(i)   temperature of reaction
(j)   speed of agitation
(k) time of addition of raw material
(l)   type of catalyst

(m) concentration of an ingredient
(n) duration of reaction
(o) stirred or not stirred
(p) particular operators.

Of the variables listed above, (i), (j), (k), (m) and (n) can be described as *quantitative* variables since they can be set at different values. Variables (l), (o) and (p) on on the other hand are described as *qualitative* variables since they cannot be given different values or even quantified at all.

An *experiment* consists of two or more *trials* and for each trial we might use a different *treatment combination*. The simplest experiment would consist of two trials with one independent variable (or treatment) being set at different *values* in each trial. For example, we might produce a batch of pigment using 3.6 litres of acid and then produce a second batch using 3.8 litres of acid, recording the brightness of the pigment in each case. The single independent variable (volume of acid) has been deliberately set at different *levels* (3.6 and 3.8) in the two trials, whilst the response (brightness) has been observed at each trial. We also speak of levels of a qualitative factor. If, for example, three different catalysts are used in an experiment which consists of nine trials, we say that the particular independent variable, catalyst type, has three levels.

The *effect* of an independent variable is the change in response which occurs when the level of the independent variable is changed. In analysing the results of an experiment we wish to estimate the effects of the independent variables which have been deliberately (or perhaps accidentally) changed in value from trial to trial. In designing an experiment we must decide:

(a) How many trials we will carry out.
(b) Which treatment combination we will use at each trial.

Many scientists and technologists would use the words 'experiment' and 'trial' rather differently to the way we have defined them. A biologist, for example, might speak of 'a large scale trial involving many experiments' whereas we will speak of 'a large scale experiment involving many trials'.

There may be no limit to the many different experiments one might design in any given situation. On the other hand there will be strict limits on the conclusions one might draw from the results of any particular experiment *after* it has been carried out. It is essential therefore to regard design and analysis as being interdependent activities. This point will be illustrated in subsequent chapters, by considering the difficulties in analysis which result from imperfect design.

9
Detecting process changes

9.1 Introduction

In later chapters of this book we will be concerned with experiments. We will discuss the advantages to be gained by carrying out a well designed experiment and we will struggle with the difficulties that arise when we try to analyse the results of a bad experiment. For the researcher who wishes to draw conclusions about cause and effect relationships in a production system the most powerful approach is that based on the designed experiment but he is unwise to completely ignore the vast amount of data which is gathered during routine production. By means of a simple statistical technique it is possible to look back at the data recorded for recent batches of product and to decide:

(a) *if* changes have taken place, and
(b) roughly *when* the changes occurred.

The technique will be illustrated by an example in which complaints have been received concerning the quality of recent batches of a pigment.

9.2 Examining the previous 50 batches

A particular pigment is manufactured by Textile Chemicals Ltd, using a batch process. This pigment has been sold in varying quantities to a variety of customers throughout Europe and the Middle East, for many years. Recently several complaints have been received from local customers concerning what they considered to be an unacceptably high level of a certain impurity in the pigment. The research and development manager of Textile Chemicals has been asked to investigate the problem and, with the co-operation of the plant manager, he has extracted from the plant records the analytical results on the last 50 batches. These include determinations of yield and impurity which are given in Table 9.1. The impurities are presented in Fig. 9.1.

The determinations of yield and impurity were both recorded to three significant figures but they have been rounded to whole numbers in Table 9.1.

Table 9.1   Yield and impurity of 50 batches of pigment

Batch number	Yield	Impurity	Batch number	Yield	Impurity
1	88	5	26	88	2
2	92	8	27	87	5
3	88	5	28	89	1
4	92	4	29	89	4
5	91	7	30	90	2
6	87	6	31	92	3
7	88	4	32	88	1
8	91	7	33	90	5
9	93	6	34	93	3
10	89	5	35	91	6
11	92	7	36	91	3
12	90	5	37	86	2
13	91	3	38	87	4
14	94	2	39	89	4
15	95	3	40	88	5
16	89	4	41	91	2
17	95	6	42	90	5
18	93	3	43	86	7
19	94	2	44	88	6
20	89	1	45	89	3
21	90	4	46	91	5
22	88	3	47	88	4
23	87	1	48	90	2
24	92	3	49	89	3
25	90	4	50	92	5

This has been done to simplify the calculations and other data used in this book will be similarly rounded. In a real life situation one would certainly not round off the data in this way.

Clearly the percentage impurity and the yield both vary from batch to batch. This variation may be quite random but it might be possible to detect patterns of change amongst the randomness. For example we might find, in either or both of the variables:

(a) An upward or a downward trend.
(b) A sudden increase or decrease (i.e. a step change).

Trends will be dealt with in later chapters but we will now concern ourselves with a statistical technique that is ideally suited to this type of situation in which we are looking back at recent data in order to detect whether or not a step-like change has occurred. By means of this technique, known as 'cusum analysis', we can produce the graphical representation of our impurity data in Fig. 9.2. ('Cusum' is an abbreviation of 'cumulative sum'.)

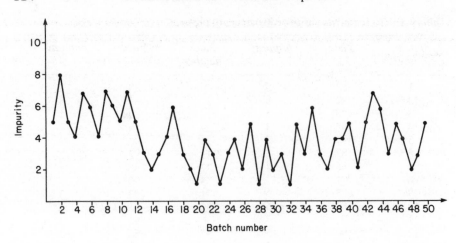

Figure 9.1   Impurity of 50 batches of pigment

When we examine Fig. 9.2 we are struck most forcibly by a very clear pattern. This same pattern is actually embedded in Fig. 9.1, but it is not so easily discernable in the simple graph. Though Fig. 9.1 and Fig. 9.2 transmit the same message, it is more easily received from the latter.

Two questions need to be answered at this point:

(a) What must we do to the data to convert Fig. 9.1 into Fig. 9.2?
(b) How are we to interpret Fig. 9.2?

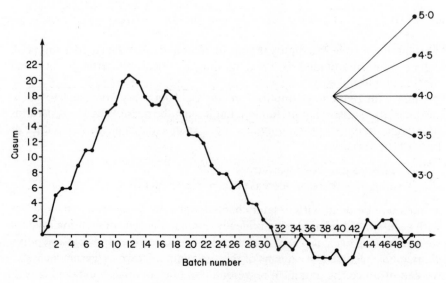

Figure 9.2   Cusum plot of impurity, with slope indicator

The calculations carried out in order to transform the impurity data of Fig. 9.1 into the cusum of Fig. 9.2 are set out in Table 9.2. Two steps are involved; first we subtract a target value from each impurity determination to obtain the deviations in column 3, then we add up these 'deviations from target' in order to obtain the 'cusum' in column 4. The cusum column is simply a running total of the deviation column. Fig. 9.2 is a graph of column 4, plotted against the batch number in column 1.

Table 9.2   Calculation of cusum for impurity data

Batch number	Impurity	Deviation from target value	Cusum	Deviation from previous batch	Squared deviation
1	5	1	1	–	–
2	8	4	5	−3	9
3	5	1	6	3	9
4	4	0	6	1	1
5	7	3	9	−3	9
6	6	2	11	1	1
7	4	0	11	2	4
8	7	3	14	−3	9
9	6	2	16	1	1
10	5	1	17	1	1
11	7	3	20	−2	4
12	5	1	21	2	4
13	3	−1	20	2	4
14	2	−2	18	1	1
15	3	−1	17	−1	1
16	4	0	17	−1	1
17	6	2	19	−2	4
18	3	−1	18	3	9
19	2	−2	16	1	1
20	1	−3	13	1	1
21	4	0	13	−3	9
22	3	−1	12	1	1
23	1	−3	9	2	4
24	3	−1	8	−2	4
25	4	0	8	−1	1
26	2	−2	6	2	4
27	5	1	7	−3	9
28	1	−3	4	4	16
29	4	0	4	−3	9
30	2	−2	2	2	4
31	3	−1	1	−1	1
32	1	−3	−2	2	4
33	5	1	−1	−4	16
34	3	−1	−2	2	4
35	6	2	0	−3	9
36	3	−1	−1	3	9
37	2	−2	−3	1	1
38	4	0	−3	−2	4
39	4	0	−3	0	0

Statistics in Research and Development

Table 9.2   Calculation of cusum for impurity data – *continued*

Batch number	Impurity	Deviation from target value	Cusum	Deviation from previous batch	Squared deviation
40	5	1	−2	−1	1
41	2	−2	−4	3	9
42	5	1	−3	−3	9
43	7	3	0	−2	4
44	6	2	2	1	1
45	3	−1	1	3	9
46	5	1	2	−2	4
47	4	0	2	1	1
48	2	−2	0	2	4
49	3	−1	−1	−1	1
50	5	1	0	−2	4
Total	200.0	0.0	–	–	230
Mean	4.0	0.0	–	–	–
SD	1.678	1.678	–	–	–

Looking back over a whole set of data in this way is often referred to as a *post-mortem investigation* but cusums are also used in quality control work when we have a set of points which grows larger as each sample is inspected. In the quality control application the target value might well be obtained from a production specification and this is discussed fully in another book of this series, *Statistical Quality Control* (Caulcutt and Boddy, 1984). *In the post-mortem investigation we use the sample mean as a target value.* As the mean impurity of the 50 batches is 4.0 we subtract 4.0 from each entry in column 2 to obtain the deviations in column 3. One consequence of using the sample mean is that the final entry in the cusum colum is zero and a second consequence is that the cusum graph (Fig. 9.2) *starts and finishes on the horizontal axis.* It is possible that a cusum plot will wander up and down in a random manner, never deviating far from the horizontal axis. It may, on the other hand, move well away from the axis only to return later. If this is the case there will be *changes in the slope* of the graph. Visual inspection of Fig. 9.2 might lead us to conclude that there are two points at which the slope changes; at batch number 12 and at batch number 32. These two 'change points' split the cusum graph (Fig. 9.2) into three sections and hence split the 50 batches into three groups:

(a) from batch 1 to batch 12 the graph has a *positive* slope;
(b) from batch 13 to batch 32 the graph has a *negative* slope;
(c) from batch 33 to batch 50 the graph is approximately horizontal (i.e. zero slope).

The slope of a section of the cusum graph is related to the mean impurity for the batches in that section. Thus the *positive* slope for batch 1 to 12 tells us that the mean impurity of these batches is *above* the target value (4.0) and the *negative* slope for batches 12 to 32 tells us that these batches have a mean impurity *below* the target value.

The radiating lines in the top right hand corner of Fig. 9.2 help us to relate the slope of a section of the cusum plot to the mean impurity of the batches in that section. Clearly a sequence of batches which has a mean impurity of 4.0 can be expected to give a section of graph which is horizontal. A change in mean impurity will give a change in slope and the steepness of the slope will depend upon the deviation of the mean from the target value of 4.0.

The changes in mean impurity which are indicated by the changes in slope in Fig. 9.2 are summarized in the stark simplicity of Fig. 9.3.

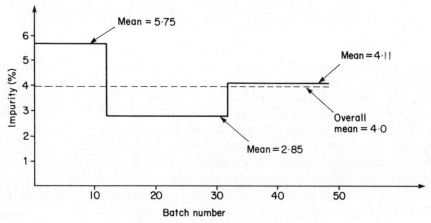

Figure 9.3   Changes in mean impurity

9.3   Significance testing

Before we translate into practical terms what we appear to have found we will examine its statistical significance. If the pattern that we have detected in the cusum graph (Fig. 9.2) is telling us anything useful then the continuous line in Fig. 9.3 should fit the data much better than does the dotted line. In other words the batch impurity values should be much closer to their local means (5.75, 2.85 and 4.11) than to the overall mean (4.0). Whether this is true or not can be judged by the summary in Table 9.3.

Table 9.3

Group of batches	Mean impurity	Standard deviation
1 to 12	5.75	1.288
13 to 32	2.85	1.387
33 to 50	4.11	1.491
1 to 50	4.00	1.784

We see in Table 9.3 that the 'within group standard deviations' (1.288, 1.387 and 1.491) are noticeably smaller than the overall standard deviation (1.784). We also notice that the three 'within group standard deviations' are approximately equal to each other indicating that the batch to batch variation is roughly the same in each group of batches. So, even if the process mean impurity *has* changed twice during this period, the process variability appears to have remained constant and we can combine the three standard deviations to obtain *one* estimate of the batch to batch variability in impurity:

Combined standard deviation

$$= \sqrt{\{\Sigma[(\text{degrees of freedom})\ (\text{within group SD})^2]/\Sigma(\text{degrees of freedom})\}}$$

$$= \sqrt{\{[(11 \times 1.288^2) + (19 \times 1.387^2) + (17 \times 1.491^2)]/[11 + 19 + 17]\}}$$

$$= 1.404$$

This combined standard deviation (1.404) is a good estimate of the batch to batch variation in impurity that we would find *if there were no changes in mean impurity* during the period in which the 50 batches were produced. An alternative method of estimating this variability is illustrated in the last two columns of Table 9.2. The entry of 230 at the bottom of the last column is the sum of the squared deviations in the penultimate column. These deviations are obtained by subtracting the impurity of each batch from that of the previous batch. This method is an alternative to using deviations from the overall mean (4.0) which would be increased by any local changes in mean which might occur. Using the 230 from Table 9.2 we can substitute into the formula:

$$\text{Localized standard deviation} = \sqrt{\left[\frac{\Sigma(x_c - x_{c+1})^2}{2(n-1)}\right]}$$

$$= \sqrt{\left[\frac{230}{2(49)}\right]}$$

$$= 1.532$$

The value of the localized standard deviation (1.532) is a little higher than the combined standard deviation (1.404) but it is much lower than the overall standard deviation (1.784), as we would expect. The localized standard deviation (1.532) is more difficult to calculate than the other two since its very nature does not allow us to make use of the standard deviation facility on a calculator. Nonetheless it is very important because we use it in testing the statistical significance of the pattern we have observed in our cusum graph. This is achieved as follows:

Null hypothesis – There was no change in process mean impurity during the period in which the 50 batches were manufactured.

Alternative hypothesis – There was a change in process mean impurity during this period.

Test statistic = $\dfrac{\text{maximum cusum}}{\text{localized standard deviation}}$

$\qquad = \dfrac{21.0}{1.532}$

$\qquad = 13.71$

Critical values – from Statistical Table I, for a span of 50 observations:

9.1 at the 5% level of significance
10.4 at the 1% level of significance.

Decision – As the test statistic is greater than the 1% critical value we reject the null hypothesis.

Conclusion – We conclude that a change of process mean impurity *did* occur during the period in question and it is reasonable to suggest that the change occurred between batches 12 and 13.

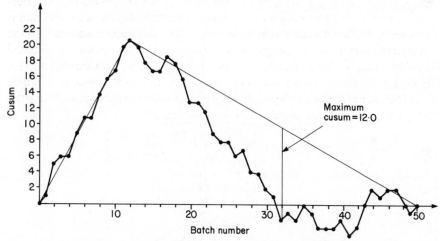

Figure 9.4   Modified cusum plot – to detect further changes

We have established beyond reasonable doubt that one of the changes in mean impurity indicated by our cusum graph (Fig. 9.2) did actually occur. We now split the whole series of 50 batches into two sets:

(a) batch numbers 1 to 12;
(b) batch numbers 13 to 50.

We could now make a fresh start and produce a cusum plot for each of the two sets of batches. This is not necessary, however, as we can obtain the same results by drawing two straight lines on the old cusum plot (Fig. 9.2) and measuring deviations from these lines. This is illustrated in Fig. 9.4.

Examination of Fig. 9.4 reveals that batch numbers 13–50 have a maximum cusum of 12 and this value can be used in a second significance test as follows:

Null hypothesis – There was no change in process mean impurity during the period in which batches 13 to 50 were manufactured.

Alternative hypothesis – There was a change in process mean during this period.

$$\text{Test statistic} = \frac{\text{maximum cusum}}{\text{localized standard deviation}}$$

$$= \frac{12.0}{1.532}$$

$$= 7.83$$

Critical values – from Table I for a span of 38 observations:

7.74 at the 5% significance level
9.04 at the 1% significance level.

Decision – As the test statistic is greater than the 5% critical value we reject the null hypothesis.

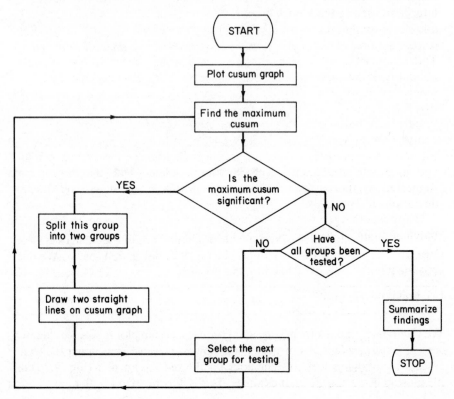

Conclusion – We conclude that a change in process mean impurity *did* occur during the period in question and it is reasonable to suggest that the change occurred about the time that batch number 32 was produced.

We have now established that *two* changes in mean occurred during the period in which batches 1 to 50 were manufactured. Further significance testing would proceed as follows:

(a) Examine the maximum cusum for batches 1 to 12.
(b) Split batches 13 to 50 into two groups (i.e. 13 to 32 and 33 to 50).
(c) Draw two more straight lines on the cusum graph.
(d) Examine the maximum cusum for batches 13 to 32.
(e) Examine the maximum cusum for batches 33 to 50.

Following through these five steps would, however, fail to reveal any more significance changes and the analysis would cease at this point. The whole procedure of splitting the batches into groups and testing each group is summarized by the flow chart.

9.4   Interpretation

Concluding that the mean impurity decreased around batch number 12 and later increased around batch number 32 is, in itself, of little use to the research and development manager. He wants to know *why* the impurity level changed so that action can be taken to ensure a low level of impurity in future batches. The establishing of cause and effect relationships can, however, be greatly helped by knowing *when* a change took place. The research and development manager can now discuss with plant personnel the events which occurred at the two points in time which have been highlighted by the cusum analysis.

Before embarking on this discussion it would be wise to carry out a cusum analysis on the *yield* data in Table 9.1. This is illustrated by the cusum plot in Fig. 9.5 which reveals two significant changes in mean yield; at batch number 7 and at batch number 19. Splitting the 50 batches into the three groups suggested by the cusum plot we get the within group means and standard deviations in Table 9.4.

It appears that the mean yield of the process increased by about 3% around batch number 7 and later decreased by approximately the same amount around

Table 9.4

Group of batches	Mean yield	Standard deviation
1 to 7	89.4	2.149
8 to 19	92.2	2.167
20 to 50	89.3	1.811
1 to 50	90.0	2.268

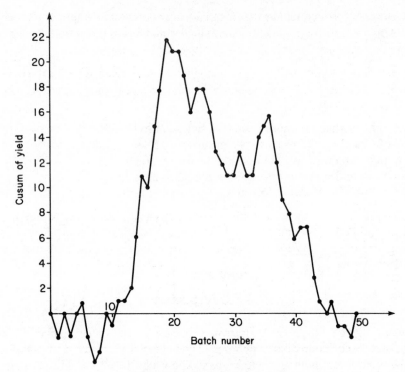

Figure 9.5   Cusum plot of yield

batch number 19. You will recall that the mean impurity decreased from 5.75% to 2.85% around batch number 12 and then increased again to 4.11% around batch number 32. All of these detected changes are illustrated in Fig. 9.6. This useful figure is often referred to as a Manhattan diagram because of its resemblance to the skyline of that part of New York City.

The research and development manager is now in a position to discuss these findings with the plant manager and other interested parties with the objective of deciding *why* these changes took place. Such discussions are not always fruitful, of course, especially if information concerning unscheduled occurrences is not revealed. After a prolonged interchange of views the following conclusions were reached:

(a) The decrease in impurity (around batch 12) was attributed to the introduction of a special ingredient which had not been used previously. This ingredient was first used in batch number 10 and was introduced with the specific aim of reducing impurity. It is not at all surprising, then, that we found a decrease in mean impurity around batch 12.

(b) No chemical explanation could be found for the increase in yield around batch number 7. It was felt by the research and development manager that the special interest shown in the plant by his chemists, at about the time

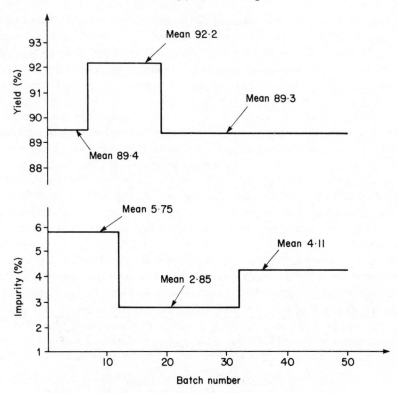

Figure 9.6 Changes in mean impurity and mean yield

that the special ingredient was introduced, may have been indirectly responsible for the increase in mean yield. The later withdrawal of research and development support could also account for the decrease in mean yield around batch number 19. It appeared that no one was aware that the mean yield was higher during this period even though an increase of 3% is regarded as quite important in profit terms.

(c) The increase in impurity around batch number 32 was the main focus of attention. This finding confirmed what had been suggested by the customer complaints, that a recent increase in impurity *had* occurred. Plant records indicated that the same weight of special ingredient had been used in every batch since the successful introduction of this additive in batch number 10. Indeed, the plant records showed that production of this pigment had proceeded very smoothly for some time.

The knowledge that mean impurity increased at around batch number 32 had focused attention on the operation of the plant during quite a short time period but no concrete explanation for the change had arisen. Thus the field was wide open for each member of the meeting to ride his personal hobby-horse and to suggest what *might* have occurred. The research and development manager had great faith in the special ingredient, the use of which had successfully reduced the impurity. He felt sure that the right quantity was not being used, though he had to admit that this very expensive substance needed to be used sparingly. The plant manager suggested that impurity could be reduced *and* yield increased if only he had the improved control systems that he needed to control the feedrate and the inlet temperature of the main ingredient. The assistant plant manager recalled that ageing of the catalyst had at one time been held responsible for excess impurity. He felt that it had been a mistake to introduce the new working arrangements which allowed the plant operators more discretion to choose *when* catalyst changes would take place. The night shift foreman yawned, and muttered that few batches of pigment would ever reach the warehouse if the operators did not use their discretion to bend the rules laid down by management. He added, in a louder voice, that there was some doubt amongst plant personnel concerning the best setting for the agitator speed.

The discussion might have continued much longer but it was brought to a conclusion when the research and development manager suggested that the most urgent need was for a better *understanding* of why the yield and impurity varied from batch to batch. It was his opinion that this understanding would only be obtained by carrying out a controlled *experiment* on the plant. The plant manager, who had anticipated this suggestion, agreed that an experiment should take place provided that it was entirely under his control.

9.5 Summary

In this chapter we have made use of one statistical technique – cusum analysis in a post-mortem setting. Though the technique is, in essence, very simple its use can nonetheless be dramatically successful in some cases. Later chapters will contain discussion of other statistical techniques which are even more powerful. Unfortunately they are not so simple as cusum analysis and the calculations may be so tedious that a computer is needed.

In the next chapter we will examine the experiment carried out by the plant manager and attempt to draw conclusions which can be translated into a practical strategy for future production. We will use a statistical technique to help us reach a decision about which of the variables included in the experiment can be used to control the level of impurity.

Problem

(1) Mr Jones always buys exactly 4 gallons of petrol at the nearest convenient garage after his petrol gauge registers half full. He always records the mileage since his last visit to a petrol station. Given below is a sequential record of the mileage recorded at the time of each purchase.

(a) Calculate the mean mileage.

(b) Calculate the cusum and plot it on a graph.
(c) Carry out significance tests to determine whether there are any different sections within the data, using a localized standard deviation of 7.83.
(d) Using the three different sections found in part (c), calculate a within group standard deviation for each section and hence compute a combined standard deviation. Compare this value with the localized standard deviation used in part (c).

Purchase	Mileage	Purchase	Mileage	Purchase	Mileage	Purchase	Mileage
1	135	11	129	21	104	31	119
2	123	12	121	22	123	32	126
3	134	13	148	23	115	33	134
4	141	14	137	24	103	34	126
5	127	15	132	25	103	35	128
6	126	16	136	26	122	36	110
7	132	17	120	27	115	37	114
8	120	18	103	28	131	38	122
9	122	19	116	29	120	39	123
10	141	20	131	30	111	40	137

———————10———————
Investigating the process – an experiment

10.1 Introduction

In Chapter 9 we used a cusum post-mortem analysis in order to discover when and why the impurity and the yield of a process had changed. By examining the plant data concerning *past* batches of pigment, we hoped to acquire some information which would help us to exercise better control over *future* batches. The exercise was not entirely successful. We detected changes in mean yield and in mean impurity but we were unable to account for all of these by reference to 'incidents' on the plant.

The cost of the cusum analysis was minimal as it *made use of existing data*; but herein lies both the strength and the weakness of the technique. The 50 batches examined were not produced in order to facilitate learning about the process. They were simply produced under 'normal' operating conditions. We may be much more successful in gaining a better understanding of the production process if we examine the results of a planned experiment.

It was with these thoughts in mind that the research and development manager suggested, in Chapter 9, that an experiment be carried out. The plant manager agreed to organize such an investigation and Chapter 10 will be devoted to analysing the results. Though the data resulting from the experiment will be more expensive to obtain than those used in the cusum analysis we can anticipate that they will be more useful to us in our search for better operating conditions.

10.2 The plant manager's experiment

You may recall that several variables were referred to in the meeting which was arranged for discussion of the cusum analysis. Each participant had a favourite variable which he felt was related to the percentage impurity and which would need to be controlled if impurity were to be reduced. The variables were:

(a) weight of special ingredient;
(b) catalyst age;
(c) feedrate of the main ingredient;

(d) inlet temperature of the main ingredient;
(e) agitation speed.

The plant manager decides to include all five as independent variables in his experiment. The two dependent variables will be impurity and yield. He also decides that the experiment will be limited to ten batches of pigment and the results of the ten trials are given in Table 10.1. (Yield values have been omitted from Table 10.1 as we will concentrate on only one of the dependent variables. In practice we would consider each of the dependent variables separately and then attempt to draw 'compromise' conclusions.) Note the double line separating the *y* column from the other columns in Table 10.1. The distinction between independent variables on the one hand, and the response or dependent variable on the other hand, is vitally important. When recommending a production strategy to the plant manager we must specify values for the independent variables and predict the level of response (i.e. impurity) which can be expected if the strategy is adopted.

Table 10.1   The plant manager's experiment

Impurity in pigment y	Weight of special ingredient x	Catalyst age w	Main ingredient		Agitation speed s
			Feedrate z	Inlet temp. t	
4	3	1	3	1	3
3	4	2	5	2	3
4	6	3	7	3	3
6	3	4	5	3	2
7	1	5	2	2	2
2	5	6	6	1	2
6	1	7	2	2	2
10	0	8	1	3	4
5	2	9	3	1	4
3	5	10	6	2	4
Mean 5.0	3.0	5.5	4.0	2.0	2.9
SD 2.36	2.00	3.03	2.05	0.82	0.88

The research and development manager asks one of his young chemists to work with the plant manager in the analysis of these data. The purpose of this analysis is to:

(a) identify those variables which could be changed in order to reduce the impurity in the pigment;
(b) ascertain the nature of the relationship between these chosen variable(s) and the percentage impurity;
(c) recommend a strategy for the production of future batches.

10.3   Selecting the first independent variable

From our five independent variables we will first *select the one which appears to be most closely related to the dependent variable.* Before we perform any calculations let us examine the scatter diagrams in Fig. 10.1, which illustrate these relationships. In each of the five diagrams the dependent variable is on the vertical axis and one of the independent variables is on the horizontal axis. Each diagram illustrates the association between the percentage impurity in the pigment and one of the variables which might be having some influence upon it.

A visual inspection of the five diagrams reveals that:

(a) Those batches which contain larger quantities of special ingredient tend to have a low level of impurity in the pigment.
(b) Those batches which were manufactured with a high feed rate tend to have a low level of pigment impurity.
(c) Use of a higher agitation speed may result in a higher level of impurity, but the association between the two variables is not so strong and the visual impression is perhaps over influenced by the batch which had 10% impurity.
(d) A high level of impurity in the pigment may be associated with the use of a high inlet temperature for the main ingredient.
(e) It is difficult to decide whether or not the dependent variable is influenced by the age of the catalyst.

The drawing of scatter diagrams, and the careful inspection of these diagrams, can often bring to light features of a set of data which might otherwise go unnoticed, and the utility of a simple scatter diagram should not be under-estimated. Despite this, there is also a need to express the *degree of association between two variables* in a numerical form. This is achieved by calculating the *covariance* and the *correlation coefficient.*

In Part One of this book we used the variance as a measure of the variability or spread in the values recorded for *one* variable. The covariance of *two* variables is a measure of how they vary *together.* It is calculated from:

$$\text{Covariance of } x \text{ and } y = \Sigma\,(x-\bar{x})\,(y-\bar{y})/(n-1)$$
$$\text{or} \quad (\Sigma\,xy - n\bar{x}\bar{y})/(n-1)$$

The use of the first formula will be illustrated by calculating the covariance of 'pigment impurity' and 'weight of special ingredient' (Table 10.2).

The covariance of the two variables, pigment impurity and weight of special ingredient, is -4.0. It has been suggested that the covariance is a measure of the degree of association between the two variables, but what exactly does it mean? Unfortunately the figure of -4.0 does not tell us a great deal. The most

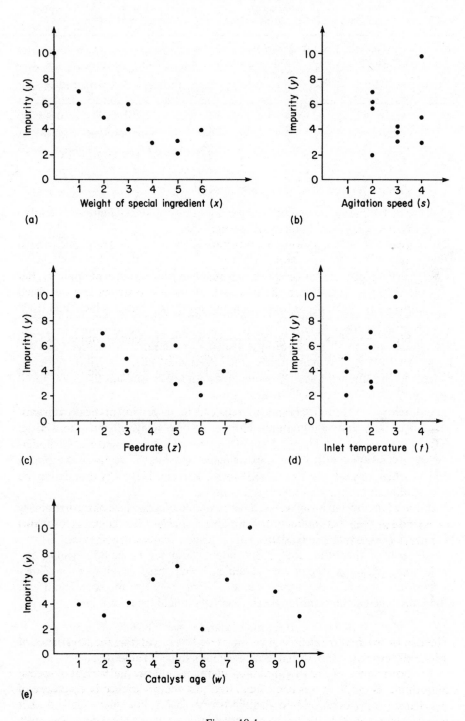

Figure 10.1

Table 10.2   Calculation of covariance

Pigment impurity	Weight of special ingredient			
y	x	$(y - \bar{y})$	$(x - \bar{x})$	$(x - \bar{x})(y - \bar{y})$
4	3	−1	0	0
3	4	−2	1	−2
4	6	−1	3	−3
6	3	1	0	0
7	1	2	−2	−4
2	5	−3	2	−6
6	1	1	−2	−2
10	0	5	−3	−15
5	2	0	−1	0
3	5	−2	2	−4
Total 50	30	0	0	−36

Covariance = −36/9 = −4.000

useful result of the calculation is the *minus sign*. This tells us that there is a negative association between the two variables, i.e. large values of x are associated with small values of y, whilst small values of x are associated with large values of y. The magnitude of the covariance (4.0), in addition to being expressed in rather unfortunate units, does not tell us whether the association between the two variables is strong or weak.

A much more useful measure of association is the correlation coefficient which is usually represented by the letter r and is calculated from:

$$\text{Correlation coefficient of } x \text{ and } y = \frac{\text{covariance of } x \text{ and } y}{(\text{SD of } x)\,(\text{SD of } y)}$$

Operating on the covariance we have just calculated, gives the correlation coefficient of pigment impurity and weight of special ingredient as (−4.000)/ (2.357 × 2.000) which is −0.8485.

For each of the other four independent variables we could calculate its covariance with the dependent variable. Using these covariances and the appropriate standard deviations we could then calculate the correlation coefficients. The calculations have been carried out and the results are:

$$r_{xy} = \text{correlation of } x \text{ and } y = -0.85$$

$$r_{wy} = \text{correlation of } w \text{ and } y = 0.23$$

$$r_{zy} = \text{correlation of } z \text{ and } y = -0.78$$

$$r_{ty} = \text{correlation of } t \text{ and } y = 0.52$$

$$r_{sy} = \text{correlation of } s \text{ and } y = 0.11$$

These correlation coefficients tell us how strongly the independent variables are associated with the dependent variable. Since the largest of the correlations is that between *x* and *y* (ignoring the minus signs) we can say that *the independent variable most strongly associated with the percentage impurity in the pigment (y) is the weight of special ingredient (x).* This statement refers only to the *sample* of ten batches and not to the population of all batches about which the research and development manager would like to draw conclusions or make

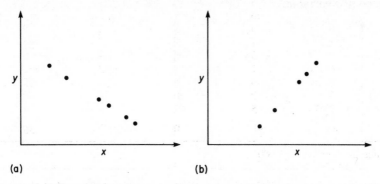

(a)   (b)

Figure 10.2   Extreme values of correlation coefficient

predictions. Before we turn our attention to the correlation coefficients that one might find if one examined all possible batches we will take a closer look at the meaning one can extract from a correlation coefficient.

It can be shown that a correlation coefficient will have a value between -1 and $+1$. A value of -1 corresponds to a perfect negative association between the two variables [Fig. 10.2(a)] whilst a value of $+1$ corresponds to a perfect positive association [Fig. 10.2(b)].

Between the two extreme cases of Fig. 10.2 we have the five scatter diagrams in Fig. 10.1. Note that Fig. 10.1(a) and Fig. 10.1(c) have negative values of correlation coefficient, whilst the weakness of the association between *w* and *y* is reflected in the correlation of 0.11 which is close to zero. A correlation of 0 is

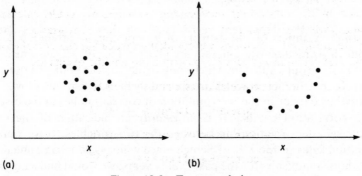

(a)   (b)

Figure 10.3   Zero correlation

an indication of the absence of a *linear association* between the two variables. It would be unwise, however, to assume that a correlation of 0 was evidence of the independence of the two variables, as Fig. 10.3(b) demonstrates.

Both parts of Fig. 10.3 illustrate sets of data which give a correlation of 0. Whilst Fig. 10.3(a) strongly suggests that the variables x and y are unrelated, Fig. 10.3(b) suggests that x and y are very closely related but the relationship is not linear. *In short, the correlation coefficient is a measure of linear association between two variables.*

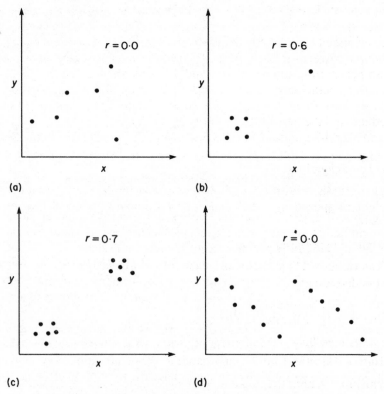

Figure 10.4   Grouping and/or outliers can affect correlation

Further examples of the dangers of interpreting a correlation coefficient without drawing a scatter diagram are given in Fig. 10.4. Data which fall into two or more groups and data which contain outliers can both give rise to values of correlation coefficient which are spuriously high or spuriously low.

A further danger in the interpretation of correlation lies in the assumption that a correlation coefficient is *necessarily* an indication of the existence of a cause–effect relationship between the two variables. If we were to list the annual sales in the UK of Scotch whisky alongside mean annual salaries of Methodist ministers for the past twenty years, we would find a high positive correlation between the two variables, but no one would suggest that there is a direct cause–effect relationship between the two. A voluntary reduction in ministers' salaries will not result in a reduction in the sales of whisky any more than a drop in the sales of whisky will cause a decrease in the ministers' salaries.

10.4 Is the correlation due to chance?

We have selected 'weight of special ingredient' (x) as the independent variable which is most closely associated with the percentage impurity in the pigment. This decision was based on the values of the five correlation coefficients. These were calculated from data gathered from the ten batches in the sample. They are *sample* correlation coefficients. Having selected this independent variable we can now turn to the population and ask 'Is it possible that the population correlation coefficient is equal to zero?' If we had examined all possible batches would we have found that there was *no* association between weight of special ingredient and the percentage impurity in the pigment?

To answer this question we carry out a simple hypothesis test. In doing so we will represent the sample correlation coefficient by r and the population correlation coefficient by ρ.

Null hypothesis – Within the population of batches the correlation between weight of special ingredient and pigment impurity is equal to zero (i.e. $\rho = 0$).

Alternative hypothesis: $\rho \neq 0$

Critical values – From Statistical Table H (with a sample size of ten, for a two-sided test):

0.632 at the 5% significance level
0.765 at the 1% significance level.

Test statistic = 0.845 (the modulus of the sample correlation coefficient, $|r|$).

Decision – Reject the null hypothesis at the 1% significance level.

Conclusion – Within the population of batches the percentage impurity in the pigment *is* related to the weight of special ingredient used in the batch.

Note: Strictly, the use of Table H is only valid if the sample is a random sample from the population referred to in the null hypothesis, but it is clearly not possible to satisfy this condition. A further assumption underlying Table H is that both x and y have normal distributions, though the table is often used in situations where this assumption is not satisfied.

10.5 Fitting the best straight line

We are now confident that the variation from batch to batch of the weight of special ingredient (x) is associated with the variation in pigment impurity (y). It is possible that there is a cause–effect relationship between the two variables and that a deliberate attempt to change the weight of special ingredient in future batches will result in a reduction in pigment impurity. We know from the *negative sign* of the correlation coefficient that an increase in x will give a reduction in y but we do not know the precise nature of the relationship. It would be useful to know what particular weight of special ingredient should be used to give a particular level of impurity. In order to quantify this relationship we will fit a straight line to the scatter diagram of Fig. 10.1(a).

If each reader of this book drew on the diagram what he/she considered to be the 'best' straight line we might well get a wide difference of opinions within a general area of agreement. One way in which we can reach complete agreement is for everyone to fit what is known as the 'least squares regression line'. This is, in one sense, the *best* straight line, but we will return to this point later after we have done the actual fitting.

If we let the equation of the line be $y = a + bx$ then fitting the line to the data simply involves calculating a value for the slope b and a value for the intercept a. This is done using:

$$\text{Slope } b = \frac{\text{covariance of } x \text{ and } y}{\text{variance of } x}$$

$$\text{Intercept } a = \bar{y} - b\bar{x}$$

Note: x represents the independent variable; y represents the dependent variable.

In our example the calculations are simple because we already have the covariance of x and y (-4.0), the standard deviation of x (2.00) and the means of the two variables.

$$\text{Slope } b = (-4.0)/(2.0)^2 = -1.0$$

$$\text{Intercept } a = 5.0 - (-1.0 \times 3.0) = 8.0$$

Thus the equation of the least squares regression line is $y = 8.0 - x$, and this equation can be used to predict the y value which corresponds to any particular x value. We can predict the percentage impurity in the pigment of future batches if we know the weight of special ingredient that is to be used in their manufacture. Substituting $x = 6$ in the equation gives $y = 2$, so we would expect batches which contain 6 units of special ingredient to have 2% impurity in the pigment on average. Our prediction would be aimed at the population of batches and not at the sample of ten batches on which the equation is based. Having studied estimation and hypothesis testing in Part One of this book you would expect that any such prediction would be in error. Later we will calculate confidence intervals for such predictions, but at this point it might help you to appreciate the need for caution in the use of a regression equation if you attempt to predict the percentage impurity that we would get if we used 9 units of special ingredient in the manufacture of a batch. Substituting $x = 9$ into the equation gives $y = -1$, but an impurity of -1% is clearly not possible.

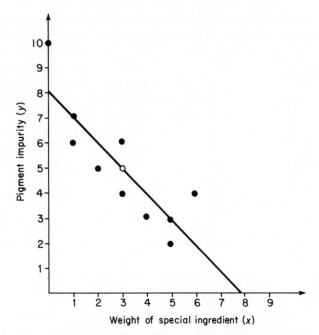

Figure 10.5   Least squares regression line

The regression line has been drawn on the scatter diagram in Fig. 10.5. We will examine this picture in some detail in an attempt to quantify 'how well the line fits the data points' and to explain how this particular line fits better than all other lines.

10.6   Goodness of fit

You will notice that the regression line passes through the point $(x = \bar{x}, y = \bar{y})$. This point is known as the *centroid* and it has been marked with a circle to distinguish it from the data points. You will also notice that two of the data points lie on the line, whilst three of the points are above the line and five are below. *The vertical distances of the points from the line are known as residuals* and they are listed in Table 10.3. Also listed are the predicted impurity values for each of the ten batches. These predicted values are obtained by substituting each x value in the regression equation.

The residual for any data point can be found from the graph or it can be calculated from:

| Residual = actual y value − predicted y value

Carrying out this calculation will give a positive residual for a point which lies above the line and a negative residual for a point which lies below. You will notice that the sum of the residuals is equal to 0 in Table 10.3. For the least

Statistics in Research and Development

Table 10.3   Calculation of residuals and the residual sum of squares

	Weight of special ingredient	Actual impurity	Predicted impurity	Residual	Squared residual
	x	y	$a + bx$		
	3	4	5	−1	1
	4	3	4	−1	1
	6	4	2	2	4
	3	6	5	1	1
	1	7	7	0	0
	5	2	3	−1	1
	1	6	7	−1	1
	0	10	8	2	4
	2	5	6	−1	1
	5	3	3	0	0
Total	30	50	50	0	14
Mean	3.0	5.0	5.0	0.0	1.4

squares regression line, or for any other line which passes through the centroid, the residuals will sum to zero. The sum of the squared residuals is equal to 14 in Table 10.3. This total is called the *residual sum of squares* and the line we have fitted is the 'best' in that it gives a smaller residual sum of squares than any other line. We have calculated values for a and b which minimize the residual sum of squares and the procedure we have used to fit the line is known as the 'method of least squares'.

The residual sum of squares will prove extremely useful when we come to calculate a confidence interval for the impurity level we can expect to get with different operating conditions. Fortunately the residual sum of squares can be calculated by other methods which avoid the tedium of tabulating the residual for each batch, as we did in Table 10.3.

> Residual sum of squares $= (n - 1) \, (\text{SD of } y)^2 \, (1 - r^2)$
>
> where y is the dependent variable and r is the correlation between the dependent and independent variables.

Using SD of $y = 2.357$, $r = -0.8485$ and $n = 10$ we get:

$$\text{Residual sum of squares} = 9(2.357)^2 \, [1 - (-0.8485)^2]$$

$$= 50.00 \, (0.2800)$$

$$= 14.00$$

This result is in agreement with that obtained earlier but the reader should note the need to carry *many* significant figures throughout the calculation. It would be unwise, for example, to use the standard deviation of y from Table 10.1 which has been rounded to two decimal places.

The residual sum of squares is a measure of the variability in the dependent variable (*y*) which has *not* been explained, or accounted for, by fitting the regression equation. Had the regression line passed through all the data points the residual sum of squares would have been equal to zero. This would be an extreme case in which the line fitted perfectly and the correlation coefficient would be equal to −1 or +1. In such a case we would say that the variability in the independent variable (*x*) accounted for 100% of the variability in the dependent variable (*y*). In our example the variation in the weight of special ingredient (*x*) from batch to batch has accounted for only *part* of the variation in pigment impurity. The ability of the fitted regression equation to explain, or account for, the variation in the *dependent variable* can be quantified by calculating a very useful figure, known as the percentage fit:

Percentage fit $= 100\, r^2$

where *r* is the correlation coefficient between the dependent variable and the independent variable.

You will recall that the correlation between *y* and *x* was −0.8485. Using this value we get a percentage fit of $100\,(-0.8485)^2$ which is 72.0%. (Note again the need to carry many significant figures in statistical calculations. Using a rounded correlation coefficient of 0.85 would give an erroneous result of 72.3% fit.)

A percentage fit of 72% tells us that the batch to batch variation in weight of special ingredient (*x*) has accounted for 72% of the variation in impurity. The remaining 28% of the variation in impurity amongst the ten batches remains unaccounted for at this point. Perhaps we will be able to demonstrate in the next chapter that this additional variation is caused, at least in part, by the other independent variables.

Another calculation which can be introduced at this point is one which converts the residual sum of squares into the very useful residual standard deviation.

Residual standard deviation (RSD) $= \sqrt{\left(\dfrac{\text{residual sum of squares}}{\text{residual degrees of freedom}}\right)}$

where residual degrees of freedom is equal to $(n-2)$ for a simple regression equation with *one* independent variable.

In Part One of this book we divided a 'sum of squares' by a 'degrees of freedom' to obtain a variance and we then took the square root to get a standard deviation. At that time we used $(n-1)$ degrees of freedom but we must now use $(n-2)$ degrees of freedom whilst calculating the residual standard deviation (RSD) of a simple regression equation. By using the calculated slope *b* and the calculated intercept *a* to obtain the residuals we lose 2 degrees of freedom compared with the *1* degree of freedom we lost earlier when we took deviations from the sample mean. Considering the list of residuals in Table 10.3, if we were given any eight of the residuals it would be possible to calculate the other two, so we have only 8 degrees of freedom amongst the ten residuals. Using a residual sum of squares of 14.0 and 8 degrees of freedom we get a residual standard deviation (RSD) equal to $\sqrt{(14.0/8)}$ which is 1.3229.

In many situations the residual standard deviation is much more meaningful than the residual sum of squares. In our pigment impurity problem the residual standard deviation is an estimate of the variation in impurity that would have been found amongst the ten experimental batches if weight of special ingredient (x) had been kept constant throughout. Note that the calculated RSD (1.323) is much less than the standard deviation of the ten batch impurity values (2.357). Obviously we can expect less batch to batch variation in impurity if weight of special ingredient (x) is held constant than if it is deliberately changed as it was in the ten-batch experiment.

10.7   The 'true' regression equation

The meaning of the residual standard deviation may be clarified further if we consider the 'true equation' or the 'population equation' as it might be called:

$$y = \alpha + \beta x \tag{10.1}$$

This equation expresses the *true* relationship between pigment impurity (y) and weight of special ingredient (x). The Greek letters alpha (α) and beta (β) remind us that the equation describes a *population* of batches and that the values of α and β could only be obtained by investigating the whole population. Furthermore, we would need to keep all other variables constant during this investigation as only one independent variable x is included in the true equation. A much better *model* of our practical situation is the equation:

$$y = \alpha + \beta x + \text{error} \tag{10.2}$$

The 'error' in this equation represents *two* causes of impurity variation which were not accounted for in equation (10.1). These two sources of variation are:

(a) Errors of measurement in the recorded values of pigment impurity (y).
(b) Batch to batch variation in impurity due to variables *other than* the weight of special ingredient (x).

Equation (10.2) tells us that the impurity of any particular batch, made with a known value of special ingredient, may well differ from the theoretical value predicted by equation (10.1). Furthermore, equation (10.2) can account for the fact that a succession of batches made with the same weight of special ingredient will be found to have differing values of impurity. It can be shown, in fact, that the residual standard deviation (1.323) is an estimate of the impurity variation we would get if the weight of special ingredient did not change from batch to batch. We have, therefore, estimated all three unknowns in equation (10.2):

(a) the calculated intercept *a* is an estimate of the true intercept α;
(b) the calculated slope *b* is an estimate of the true slope β;
(c) the residual standard deviation (RSD) is an estimate of the standard deviation (σ) of the 'errors'.

Rather than single value estimates we could calculate confidence intervals for α, β and σ. Before we do so, however, let us pose a very important question: 'Could β be equal to zero?' In practical terms this is equivalent to asking 'if we had examined a very large number of batches would we have found that the pigment impurity (*y*) *was not* related to the weight of special ingredient (*x*)?' To answer this question we will carry out a *t*-test in which the null hypothesis is $\beta = 0$.

Null hypothesis – There is no relationship between weight of special ingredient (*x*) and pigment impurity (*y*) (i.e. $\beta = 0$).

Alternative hypothesis – There is a relationship between weight of special ingredient (*x*) and pigment impurity (*y*) (i.e. $\beta \neq 0$).

$$\text{Test statistic} = \frac{|b - \beta|(\text{SD of } x)}{\text{RSD}/\sqrt{(n-1)}} \quad \text{or} \quad \sqrt{\left[\frac{\% \text{ fit } (n-2)}{100 - \% \text{ fit}}\right]}$$

$$= \frac{|-1.000 - 0.000|(2.000)}{1.3229/\sqrt{9}} \qquad = \sqrt{\left[\frac{72\,(8)}{100 - 72}\right]}$$

$$= 4.5355 \qquad\qquad\qquad = 4.5356$$

Critical values – From the two-sided *t*-table with 8 degrees of freedom:

2.31 at the 5% significance level
3.36 at the 1% significance level.

Decision – We reject the null hypothesis at the 1% level of significance.

Conclusion – We conclude that, within the population of batches, pigment impurity (*y*) *is* related to the weight of special ingredient (*x*).

We did, of course, reach this very same conclusion earlier when we tested the significance of the correlation between *x* and *y*. Why, you might ask, should we carry out the *t*-test when we already know the answer to the question? The *t*-test, as you will see in the next chapter, is easily extended to cover equations which contain two or more independent variables. The correlation test can also be extended but not so easily.

For similar reasons we have offered two formulae with which to calculate the test statistic in the *t*-test. Whilst the first formula may appear more meaningful the second formula, based on the percentage fit, can be easily modified for use in situations where we have several independent variables in the equation.

10.8   Accuracy of prediction

Having decided that the true slope β is not equal to zero it is reasonable to ask for a more positive statement about what value the true slope is likely to have. We have calculated that b is equal to -1.0 and this is our best estimate of β. We can, however, calculate a confidence interval as follows:

> A confidence interval for the true slope (β) is given by:
>
> $$ b \pm t(\text{RSD}) \sqrt{\left[\frac{1}{(n-1)\,\text{Var}\,(x)} \right]} $$
>
> where t has $(n-2)$ degrees of freedom.

Using $b = -1.0$, $t = 2.31$, RSD $= 1.323$, SD of $x = 2.00$ and $n = 10$ we get a 95% confidence interval for the true slope to be:

$$ -1.0 \pm 2.31\,(1.323) \sqrt{\left[\frac{1}{9(2.0)^2} \right]} = -1.0 \pm 0.51 $$

$$ = -0.49 \text{ to } -1.51 $$

We can, therefore, be 95% confident that the true slope lies between -0.49 and -1.51. In practical terms, this means that we can expect that an increase of 1 unit in the weight of special ingredient will result in an impurity decrease of between 0.49% and 1.51%. This is a very wide interval and the plant manager is left in some doubt. To achieve a 3% reduction in impurity does he need to increase the weight of special ingredient by approximately 2 units (i.e. 3/1.51) or by approximately 6 units (i.e. 3/0.49)? It would be easier for the plant manager to estimate the cost of a change in operating conditions if he knew more precisely the effect of such a change.

We can also calculate a confidence interval for the true intercept α by means of the formula:

> A confidence interval for the true intercept α is given by:
>
> $$ a \pm t(\text{RSD}) \sqrt{\left[\frac{1}{n} + \frac{\bar{x}^2}{(n-1)\,\text{Var}\,(x)} \right]} $$
>
> where t has $(n-2)$ degrees of freedom.

Using $a = 8.0$, $t = 2.31$, RSD $= 1.323$, $\bar{x} = 3.0$, SD of $x = 2.00$ and $n = 10$ we get a 95% confidence interval for the true intercept to be:

$$ 8.0 \pm 2.31\,(1.323) \sqrt{\left[\frac{1}{10} + \frac{(3.0)^2}{9(2.0)^2} \right]} = 8.0 \pm 1.80 $$

We can be 95% confident that the true intercept α lies between 6.20 and 9.80. Thus we can advise the plant manger that he can expect an average impurity between 6.2% and 9.8% if he does not include any special ingredient (i.e. $x = 0$) in future batches. Once again the confidence interval is disappointingly wide, but this particular estimate is not very important as the plant manager fully intends to use the special ingredient (x) in future batches. What he would appreciate, much more than a confidence interval for α, is a confidence interval for the mean impurity in future batches for a particular level of special ingredient.

A confidence interval for the true value of y for a particular value of x (say x = X) is given by:

$$a + bX \pm t(\text{RSD}) \sqrt{\left[\frac{1}{n} + \frac{(X - \bar{x})^2}{(n - 1)\,\text{Var}\,(x)}\right]}$$

where t has $(n - 2)$ degrees of freedom.

Suppose that the plant manager intends to use 6 units of special ingredient in future batches. By substituting $X = 6$ into the formula we get a confidence interval for the true impurity (or mean impurity) as follows:

$$8.0 - 1.0\,(6.0) \pm 2.31\,(1.323) \quad \frac{1}{10} + \frac{(6.0 - 3.0)^2}{9(2.0)^2}$$

$$= 2.00 \pm 1.80$$

Thus the use of 6 units of special ingredient can be expected to result in a mean impurity which lies between 0.20% and 3.80%. Whilst the manager would be very happy with the 0.2% impurity at the lower end of this range he would not be at all pleased to find 3.8% impurity if he had incurred the high cost of using 6 units of special ingredient. Using other values of x we could calculate several confidence intervals and then plot the confidence bands shown in Fig. 10.6.

The plant manger's position is actually even weaker than Fig. 10.6 would suggest. He must bear in mind that the confidence intervals in Fig. 10.6 are for the *true mean* impurity of future batches in which 6 units of special ingredient are used, and that individual batches will be scattered above and below the true mean. If we want a confidence interval for the pigment impurity in a *particular* batch (the next batch to be produced, say) we must use the following formula:

$$a + bX \pm t_{n-2}\,\text{RSD} \sqrt{\left[1 + \frac{1}{n} + \frac{(X - \bar{x})^2}{(n - 1)\,\text{Var}\,(x)}\right]}$$

Comparing this formula with the previous one we note the extra 1 under the square root sign. This gives a wider interval and if we again substitute $X = 6$ we get a 95% confidence interval of 2 ± 3.54. The plant manager can therefore

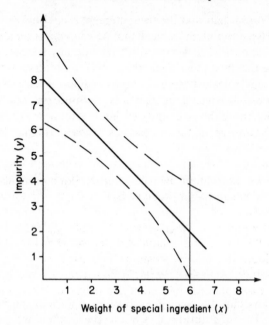

Figure 10.6   Confidence bands for predicted impurity of future batches

expect that the use of 6 units of special ingredient in the next batch will give an impurity level between -1.54% and 5.54%. The lower confidence limit is clearly impossible whilst the upper limit is alarming.

The plant manager would appear to have got little benefit from the use of regression analysis in this chapter. We have fitted a regression equation ($y = 8.00 - x$) which he himself could have 'fitted by eye' using Fig. 10.1(a). It is true that he might have ended up with an equation which differed slightly from our 'best' equation, but this error would have been negligible in the light of the confidence intervals we have just calculated. Fortunately, regression analysis has much more to offer than has so far been revealed. In the next chapter we will take into account the other four independent variables. The effect will be to increase the percentage fit of the equation and consequently reduce the residual standard deviation. This in turn will give a much narrower confidence interval for the predicted impurity of any production strategy. Before we consider the introduction of additional independent variables we will briefly examine an alternative to the simple equation ($y = 8.0 - x$) that we have already fitted.

10.9   An alternative equation

You may recall that we selected weight of special ingredient (x) as our first independent variable because of its high correlation (-0.85) with the dependent variable, pigment impurity (y). We later found this sample correlation coefficient to be significant at the 1% level. Amongst the other independent variables the only serious rival for inclusion in the equation was the feedrate (z). The correlation between z and y is -0.78 which, when tested, is found to be significant at the 5% level.

Had we decided to select feedrate (z) as our first independent variable the fitted equation would have been:

$$y = 8.6 - 0.90z$$

and the percentage fit would have been 61%. Clearly the use of z as an independent variable does not give as good a percentage fit as we got when we used x (72%). The difference is not very large, however, and this alternative equation may be far more attractive in terms of plant operation. The original equation offered a reduction in impurity of 1% for a unit increase in the weight of special ingredient (x). The alternative equation offers a reduction in impurity of 0.9% for a unit increase in the feedrate (z). When the plant manager considers the relative costs of increasing the weight of special ingredient and of increasing the feedrate he may conclude that the latter is more desirable. He may contemplate making an increase in x *and also* making an increase in z. In the next chapter we will explore this possibility.

> *A warning*
> It would be very unwise of the plant manager to change *either* the weight of special ingredient (x) *or* the feedrate *or* both, until he has studied multiple regression and the design of experiments which are dealt with in later chapters.

10.10 Summary

In this chapter we have examined the least squares method of fitting simple regression equations. The equations were used to predict the level of impurity that might result from using specified weights of special ingredient (x) and specified levels of feedrate (z). Confidence intervals were calculated for the predicted impurity using the residual standard deviation. These confidence intervals were rather wide but we will see in the next chapter that the use of two independent variables will give a *higher* percentage fit and consequently a *lower* residual standard deviation with correspondingly *narrower* confidence intervals.

Problems

(1) Edlington Chemicals produce monocylate using a batch process. The plant manager is concerned about high reaction times of certain batches and a detailed examination of past records convinces him that this is due to high levels of hexanol impurity in the feedstock. For statistical analysis he chooses a sequence of twelve previous batches in which the results look 'promising'.

Concentration of hexanol	Reaction time
10	300
13	380
10	350
11	320
7	280
14	400
9	330
13	370
10	330
9	350
12	310
14	360

(a) Decide which is the dependent variable (y) and which is the independent variable (x).

(b) Calculate $SD(x)$, $SD(y)$ and covariance (x and y).

(c) Calculate the correlation coefficient between reaction time and concentration.

(d) Test the significance of the correlation coefficient (consider carefully whether it is a one-sided or two-sided test).

(e) Calculate the slope and intercept of the regression line, $y = a + bx$.

(f) Plot a scatter diagram, with the independent variable on the horizontal axis, and draw your regression line on the diagram.

(g) Calculate the percentage fit for the regression equation.

(h) Calculate the residual standard deviation.

(i) Test the statistical significance of the relationship between the two variables using a t-test.

(j) Calculate 95% confidence intervals for the true intercept and true slope.

(k) Calculate 95% confidence intervals for the true mean reaction time of batches with concentrations of:

  (i) 7;
  (ii) 11.

(l) Calculate 95% confidence intervals for the reaction time of a single batch with concentrations of:

  (i) 7;
  (ii) 11.

(m) Has the plant manager carried out an investigation which is in accord with scientific method?

(2) Yorkshire Spinners Ltd buy triacetate polymer and convert it into different types of fabric. This process involves many stages, all of which affect the quality

of the final product, but it has been shown that the critical stage is the first one in which polymer is heated, forced through minute orifices and wound onto bobbins as a fibre. The quality of the fibre may be represented by many parameters including birefringence, which is a measure of orientation of polymer molecules in the fibre and can be controlled by changing the speed of winding onto the bobbin – referred to as wind-up speed. The wind-up speed can be adjusted quickly but other associated adjustments – flowrate and temperature – cannot.

These adjustments, necessary to keep parameters other than birefringence constant, lead to a loss of production. It is necessary that the 'correct' adjustment be made to birefringence since over- or under-correction will lead to a further loss of production.

Making the 'correct' adjustment depends upon knowing accurately the relationship between birefringence and wind-up speed. This relationship is, however, dependent upon the type of machine.

A new machine is being developed by Yorkshire Spinners and the research manager has, absent-mindedly, independently asked four scientists – Addy, Bolam, Cooper and Dawson – to evaluate the relationship between birefringence and wind-up speed. He particularly wants two results:

(a) An estimate of the slope which will be used in conjunction with hourly quality control checks on birefringence. If a change in birefringence is necessary the 'slope' will be used to indicate the change in wind-up speed.
(b) An estimate of the true relationship between the two variables to be used in the calculation of 'start-up' conditions for a new product.

The four scientists work in isolation from each other and produce different experimental designs to tackle the same problem. The designs and the experimental results are given below:

Addy

Wind-up speed	150	160	170	180	190	200
Birefringence	70.1	75.3	77.0	80.6	87.2	89.8

Bolam

Wind-up speed	150	175	175	175	175	200
Birefringence	70.0	81.3	78.1	79.9	80.5	90.2

Cooper

Wind-up speed	150	150	150	200	200	200
Birefringence	70.4	70.8	69.0	91.5	88.9	89.4

Dawson

Wind-up speed	165	169	173	177	181	185
Birefringence	76.2	78.9	78.1	79.5	83.5	83.8

Summary statistics of the four experiments are as follows, where x is wind-up speed and y is birefringence:

	$\bar{x}$	$\bar{y}$	SD (x)	SD (y)	$\Sigma\, xy$
Addy	175	80	18.708	7.448	84689.0
Bolam	175	80	15.811	6.475	84505.0
Cooper	175	80	27.386	10.933	85490.0
Dawson	175	80	7.483	3.040	84106.4

To assist computation use the formulae:

$$\text{Covariance } (x, y) = (\Sigma\, xy - n\bar{x}\bar{y})/(n-1)$$
$$\text{Residual SD} = \sqrt{\{(n-1)\,[s_y^2 - b\,\text{Cov}\,(xy)]/(n-2)\}}$$

Complete the following table by calculating:

(a) the correlation coefficients for Addy and Bolam;
(b) the least squares regression line for Addy and Bolam;
(c) the residual standard deviation and hence the 95% confidence interval of the slope for Addy and Bolam;
(d) the 95% confidence intervals for the true value of birefringence at wind-up speeds of 150, 175 and 200 for Addy. Plot the intervals given in the table for *all* five conditions and four experiments on the scatter diagrams (Fig. 10.7). Join the points to give 95% confidence bands.

	Addy	Bolam	Cooper	Dawson
Covariance			298.0	21.28
Correlation coefficient			0.995	0.935
Intercept			10.5	13.5
Slope			0.397	0.380
Residual standard deviation			1.24	1.20
95% confidence interval for slope			±0.056	±0.199
95% CI for true line at:				
150		69.9 ± 2.7	70.2 ± 2.0	70.5 ± 5.2
160	74.1 ± 1.9	73.9 ± 1.9	74.1 ± 1.6	74.3 ± 3.3
175		80.0 ± 1.3	80.0 ± 1.4	80.0 ± 1.4
190	85.9 ± 1.9	86.1 ± 1.9	85.9 ± 1.6	85.7 ± 3.3
200		90.1 ± 2.7	89.8 ± 2.0	89.5 ± 5.2

(e) Would you have expected the slope coefficients and the residual standard deviations to be less variable between the experiments?

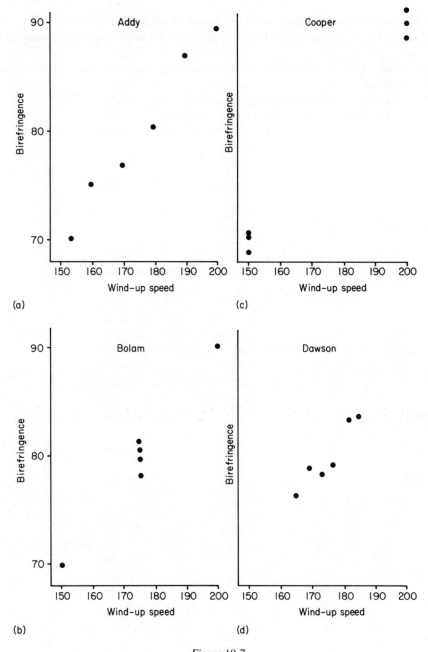

Figure 10.7

(f) Why are the confidence intervals for the slope markedly different?
(g) The four correlation coefficients are different. Is this due only to sampling error?
(h) Why is there more curvature on the confidence band for Bolam than Cooper?
(i) Is there any danger in using Dawson's design?
(j) Is there any danger in using Cooper's design?

──────11──────
Why was the experiment not successful?

11.1   Introduction

The previous chapter was centred around a situation in which we attempted to explain why the percentage impurity in a pigment varied from batch to batch. On each of ten batches several variables had been measured and we selected the 'weight of special ingredient' (x) as the independent variable which was most highly correlated with the dependent variable (percentage impurity). After fitting a simple regression equation we were able to make predictions of the impurity that could be expected in future batches if certain weights of special ingredient were used. Unfortunately the confidence intervals associated with these predictions were rather wide, which raised doubts about the usefulness of the regression equation.

We fitted a second simple equation in which the feedrate (z) was the independent variable. Whilst this new equation had a lower percentage fit (61% compared with 72%) and consequently even wider confidence intervals, it might nonetheless be more attractive to the plant manager on economic grounds. Perhaps it would be cheaper to increase the feedrate than to increase the weight of special ingredient in each batch.

In this chapter we will examine regression equations in which there are two or more independent variables. These equations will give a higher percentage fit than the simple equations discussed in Chapter 10 and we can therefore look forward to narrower confidence intervals. We will start by fitting a regression equation with *two* independent variables and it seems reasonable that we should choose feedrate (z) and weight of special ingredient (x) from the many possibilities open to us.

11.2   An equation with two independent variables

The two simple regression equations fitted in the previous chapter were:

$$y = 8.0 - x \qquad (72\% \text{ fit})$$
$$y = 8.6 - 0.90z \qquad (61\% \text{ fit})$$

We will now fit an equation which contains *both* x and z as independent variables. This will have the form:

$$y = a + bx + cz$$

and values of a, b and c will be calculated using the y, x and z measurements in Table 10.1. It is reasonable to expect that this multiple regression equation will give a higher percentage fit than either of the simple equations. On the other hand we would be foolish to expect 61% + 72% (i.e. 133%) as this would exceed the obvious upper limit of 100%.

Fitting of a multiple regression equation involves rather tedious calculations and would usually be carried out on a computer, especially if the equation contained three or more independent variables. We will, nonetheless, follow through the calculations with two independent variables as the exercise will serve to illustrate some very important principles. Using once again the method of least squares the required values of a, b c are obtained by solving the equations:

$$b \text{Var}(x) + c \text{Cov}(xz) = \text{Cov}(xy)$$
$$b \text{Cov}(xz) + c \text{Var}(z) = \text{Cov}(zy)$$
$$a = \bar{y} - b\bar{x} - c\bar{z}$$

Using the y, x and z columns of Table 10.1 we can calculate the variances, covariances and means to be:

Var (x) = 4.00   Var (z) = 4.22

Cov (xz) = 4.00   Cov (xy) = −4.0   Cov (zy) = −3.78

$\bar{x}$ = 3.0 $\bar{y}$ = 5.0 $\bar{z}$ = 4.0

Substituting the variances and covariances into the first two of the least squares equations gives:

$$4.00b + 4.00c = -4.00$$
$$4.00b + 4.22c = -3.78$$

and the solutions are b = −2.0 and c = 1.0.

Substituting into the third of the least squares equations gives:

$$a = 5.0 - (-2.0 \times 3.0) - (1.0 \times 4.0)$$

$$a = 7.0$$

The least squares multiple regression equation containing weight of special ingredient (x) and feedrate (z) as independent variables is, therefore:

$$y = 7.0 - 2.0x + 1.0z$$

What does this equation tell us about the values of x and z that the plant manager should use in order to reduce the impurity in future batches of pigment? The coefficient of x is -2.0 so we can expect that an increase of 1 unit in the weight of special ingredient (x) will give a 2% reduction in impurity. This differs from the 1% reduction offered by the simple equation ($y = 8.0 - x$). The coefficient of z in the multiple regression equation suggests that a unit increase in the feedrate (z)) will give a 1% *increase* in impurity. This is completely at odds with the 0.9% *decrease* promised by the simple equation ($y = 8.6 - 0.9z$).

The three equations that we have fitted would seem to offer very conflicting recommendations concerning the operating conditions that should be used to reduce impurity. We are not at all sure just how large a decrease in impurity could be expected from an increase in the weight of special ingredient (x). As far as the other independent variable (z) is concerned we are not even sure whether an increase *or* a decrease in impurity is likely to result from an increase in feedrate.

It is possible that *any one* of the three equations may be offering a useful indication to the plant manager but it is obviously impossible to reconcile the conflicting advice which results when we put all three equations together.

It is not unreasonable to suggest, therefore, that the use of regression analysis with this set of data has been very unsatisfactory. In fairness to the statistical technique, however, it must be pointed out that the reason for our lack of success lies not in the technique, but in the data that we have attempted to analyse. Multiple regression analysis is, without doubt, a very useful technique but great care must be exercised because certain characteristics in a set of data can result in the drawing of invalid conclusions.

> Before we attempt to use multiple regression analysis we must examine the intercorrelation of the independent variables.

You may have imagined that the five independent variables, x, w, z, t and s vary from batch to batch *independently* of each other. This is certainly not so, as we can see from the correlation matrix of Table 11.1.

Table 11.1   Correlation matrix

	x	w	z	t	s
x	1.00	−0.26	0.97	−0.07	0.00
w	−0.26	1.00	−0.20	−0.05	0.44
z	0.97	−0.20	1.00	0.07	−0.06
t	−0.07	−0.05	0.07	1.00	0.00
s	0.00	0.44	−0.06	0.00	1.00

We see from Table 11.1 that the correlation between x and z is 0.97. This implies that those batches in which a high feedrate (z) was used also had a large quantity of special ingredient (x) and vice versa. Reference to Table 10.1

confirms that this is indeed the case and we see, for example, that the third batch had $x = 6, z = 7$, whilst at the other extreme the eighth batch had $x = 0$, $z = 1$.

This high correlation between the two independent variables, x and z, is telling us that the plant manager, in producing these ten batches, was carrying out a *bad* experiment. It is this same correlation which would have warned us of the danger of using regression analysis with this data had we *first* examined the correlation matrix. Furthermore it is this same correlation (0.97) which has caused the coefficients of x and z in the multiple regression equation to differ greatly from the coefficients in the two simple equations. Indeed, when using multiple regression analysis, it is wise to watch out for large changes in coefficients as new variables enter the equation. Such changes are an indication of correlation between independent variables.

Intercorrelation of the independent variables is entirely detrimental. We have seen the disadvantages which follow from the high correlation between feedrate and weight of special ingredient. There are no compensating advantages. The correlation between x and z even intrudes when we attempt to calculate the percentage fit of the multiple regression equation.

> When designing an experiment one objective is to reduce the correlation between pairs of independent variables.

In the previous chapter we calculated the percentage fit of a simple regression equation using:

$$\text{Percentage fit} = 100r^2 \qquad (11.1)$$

To obtain the percentage fit of an equation with two independent variables we use the more complex expression:

$$\text{Percentage fit} = 100(r_{xy}^2 + r_{zy}^2 - 2r_{xy}r_{zy}r_{xz})/(1 - r_{xz}^2) \qquad (11.2)$$

As there are three correlation coefficients involved we use subscripts to distinguish between them. Thus r_{xy} represents the correlation between x and y, etc. Substituting into this equation we get:

$$\begin{aligned}
\text{Percentage fit} &= [100(-0.85)^2 + (-0.78)^2 - 2(-0.85)(-0.78)(0.97)]/ \\
&\quad [1 - (0.97)^2] \\
&= 100(0.723 + 0.608 - 1.286)/(0.059) \\
&= 76\%
\end{aligned}$$

The percentage fit of the multiple regression equation (76%) is little higher than the percentage fit of the two simple equations (72% and 61%). Once again we can blame the correlation between x and z. The effect of intercorrelation between the independent variables is highlighted if we let r_{xz} equal 0 in equation (11.2), which then becomes:

$$\text{Percentage fit} = 100(r_{xy}^2 + r_{zy}^2) \qquad (11.3)$$

Equation (11.3), which is very similar in form to equation (11.1), would give us the percentage fit if x and z were *not* correlated. In other words equation (11.3) would be applicable if the plant manager had carried out an experiment in which feedrate (z) and weight of special ingredient (x) did *not* vary from batch to batch in sympathy with each other.

11.3   Multiple regression analysis on a computer

It was stated earlier that multiple regression analysis would not usually be attempted without the support of a computer together with a suitable computer program. Perhaps it has occurred to you that the use of computing facilities could offer a safeguard against drawing erroneous conclusions from data in which the independent variables are intercorrelated. Such faith in computers can be misplaced. Certainly *some* regression analysis programs (or 'packages' as they are often called) *do* print out warnings of potential disasters, but many others *do not*. Even if you avail yourself of an excellent package it can only *warn* you that the data are suspect; it cannot turn a bad experiment into a good one. We will see in later chapters that the plant manager's experiment can be greatly improved but the improvement comes from producing several more batches and not from improving the method of data analysis.

Just as computers come in many different sizes, ranging from 'micros' through 'minis' to 'main frames', so multiple regression packages have different levels of complexity. The more sophisticated packages tend to be located in the larger machines but recent advances in computer hardware have made available very powerful multiple regression facilities on relatively inexpensive desk top computers. Just what do these packages offer?

We will answer this question by describing how one particular regression package would analyse the data from our ten batches of pigment and the print-out from this package is reproduced in Appendix D. An important feature, which this package shares with many others, is 'automatic variable selection' by means of which the program selects one independent variable, then a second, then a third, etc. Thus a sequence of regression equations is printed out until the 'best' equation is reached and then the sequence is terminated. Each new variable is selected on purely *statistical* grounds, of course, the program having no knowledge whatsoever of the underlying science/technology. Using this statistical criterion the following equations are fitted:

$$y = 8.0 - 1.0x \qquad\qquad 72\% \text{ fit}$$

$$y = 5.2 - 0.96x + 1.34t \qquad 93.4\% \text{ fit}$$

When selecting the first independent variable the program chooses the one which gives the *greatest percentage fit*. This procedure is equivalent to the one we adopted in the previous chapter when we selected weight of special ingredient (x) because it had the greatest correlation coefficient.

The inlet temperature (t) is selected as the second independent variable because it offers the *greatest increase in percentage fit*. With x already in the equation we have four variables (z, t, s and w) from which to choose. The computer program carries out calculations which show that x and t together will give a greater percentage fit than either x and z, x and s, or x and w. We will not pursue these calculations, but the wisdom of the decision can be seen if we examine Fig. 11.1

Fig. 11.1 is very similar to Fig. 10.1. The only difference between the two diagrams is on the vertical axis; previously we used impurity (y) and now we are using 'residuals'. This is a reasonable change to make. It is important to realize that by introducing the second independent variable into the regression equation we are not attempting to explain *all* the batch to batch variation in impurity (y). The first equation ($y = 8 - x$) has already accounted for 72% of the variation in y and we are now trying to explain the remaining 28%. In other words we simply wish to explain the batch to batch variation in the residuals and we are obviously very interested in any independent variable which appears to be related to these residuals. Fig. 11.1(d) indicates that the inlet temperature (t) is such a variable and this impression is confirmed if we examine the correlation of the residuals with each independent variable. These are known as *part* correlation coefficients and are given in Table 11.2.

Table 11.2   Correlation with the residuals from the first equation

Independent variable	Weight of special ingredient	Catalyst age	Main ingredient		Agitation speed
			Feedrate	Inlet temp.	
	x	w	z	t	s
Correlation with residuals	0.00	0.03	0.09	0.87	0.20

The largest entry in Table 11.2 (0.87) is the correlation between the inlet temperature (t) and the residuals from the first equation. This is known as the 'part correlation between t and y after adjusting for x' and its magnitude confirms our earlier impression that t should be the second independent variable. It seems reasonable to suggest that the larger the correlation of the new variable with the residuals the greater will be the increase in percentage fit resulting from the introduction of this variable. Whilst this is true in one sense, it can in some cases be misleading and it is therefore safer to use the *partial* correlation coefficients in Table 11.3. (A full description of part and partial correlation has been relegated to Appendix E to avoid digression at this point.)

The partial correlation coefficients in Table 11.3 have been taken from the computer print-out in Appendix D. Their usefulness is indicated by the equation below which can be used to calculate the increase in percentage fit that can be obtained by introducing any particular independent variable:

$$\text{Increase in \% fit} = (100 - \text{old \% fit})(\text{partial correlation})^2$$

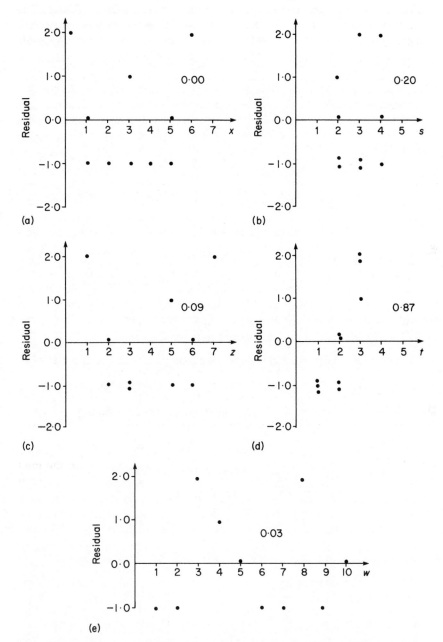

Figure 11.1   Residuals from $y = a + bx$ plotted against the independent variables

Table 11.3   Partial correlation and increase in percentage fit

Independent variable	Weight of special ingredient x	Catalyst age w	Main ingredient		Agitation speed s
			Feedrate z	Inlet temp. t	
Partial correlation with the dependent variable (x being fixed)	0.000	0.030	0.377	0.874	0.203
Increase in % fit by introducing this variable	0.0%	0.02%	4.0%	21.4%	1.2%

By introducing inlet temperature (t) into the regression equation we can expect to get an increase in percentage fit equal to:

$$(100 - 72)(0.874)^2 = 21.4\%$$

After t has been introduced as the second independent variable the percentage fit should have increased therefore from 72% to 93.4%. This is indeed the case.

Clearly the increase in percentage fit must be statistically significant or the second regression equation would not have been printed out by the computer program. We can, however, check this significance by means of a t-test in which the test statistic is based on the increase in percentage fit.

Null hypothesis – Impurity (y) is not dependent on inlet temperature (t).

Alternative hypothesis – Impurity is dependent on inlet temperature (t).

Test statistic $= \sqrt{\{[(\text{new } \% \text{ fit} - \text{old } \% \text{ fit}) \, (\text{degrees of freedom})]/}$
$(100\% - \text{new } \% \text{ fit})\}$

(Degrees of freedom $= n - k - 1 = 10 - 2 - 1 = 7$, where k is the number of independent variables in the equation.)

$$= \sqrt{\left[\frac{(93.4\% - 72\%) \, (7)}{(100\% - 93.4\%)}\right]}$$

$$= 4.76$$

Critical values – From the two-sided t-table with 7 degrees of freedom:

2.36 at 5% significance level
3.50 at 1% significance level.

Decision – We reject the null hypothesis at the 1% level of significance.

Conclusion – We conclude that impurity (y) is related to the inlet temperature.

This significance test is based on the knowledge that a relationship between impurity (y) and weight of special ingredient (x) has *already been established*. We are now able to conclude that *both* the variation in inlet temperature (t) *and* the variation in weight of special ingredient (*x*) are contributing to the batch to batch variation in impurity. If the plant manager accepts this conclusion then any strategy put forward for the reduction of impurity in future batches should specify levels for the weight of special ingredient (*x*) *and* for the inlet temperature (*t*).

Also printed out by our computer program is a table containing residuals, predicted impurity values and confidence intervals for true impurity values. This very useful table is given as Table 11.4.

Table 11.4   Residuals from the second regression equation

Weight of special ingredient	Inlet temp.	Actual impurity	Predicted impurity	Confidence interval for true impurity	Residual
x	*t*	*y*			
3	1	4	3.66	2.82 to 4.50	0.34
4	2	3	4.04	3.46 to 4.62	−1.04
6	3	4	3.45	2.25 to 4.65	0.55
3	3	6	6.34	5.50 to 7.18	−0.34
1	2	7	6.93	6.18 to 7.67	0.07
5	1	2	1.73	0.76 to 2.71	0.27
1	2	6	6.93	6.18 to 7.67	−0.93
0	3	10	9.23	8.09 to 10.36	0.77
2	1	5	4.62	3.73 to 5.52	0.38
5	2	3	3.07	2.33 to 3.82	−0.07

In Table 11.4 the column of residuals has a total of −0.04. Had there been no rounding errors in the individual residuals this total would have been zero. The sum of the *squared* residuals is equal to 3.30.

The confidence intervals in Table 11.4 vary in width. The narrowest interval has a width of 1.16 (the batch with $x = 4$ and $t = 2$) whilst the widest interval has a width of 2.4 (the batch with $x = 5$ and $t = 3$). This variation is consistent with what we found when we plotted confidence intervals from the first regression equation in Fig. 10.6. If the values of the independent variables (*x* and *t*) are close to the means ($\bar{x} = 3$ and $\bar{t} = 2$) then the confidence interval will be narrower than it would be if the values of *x* and *t* were far removed from the means.

Before we leave the computer print-out in Appendix D let us summarize the conclusions that it suggests and record our reservations:

(a) Using a statistical criterion at each stage the computer program has given us the 'best' equation:

$$y = 5.2 - 0.96x + 1.34t \qquad 93.4\% \text{ fit}$$

(b) This equation gives predicted impurity values which have much narrower confidence intervals than those which were given by the first regression equation.

(c) The correlation matrix which is included in the print-out contains one very high value (0.97) which must not be ignored. This correlation between weight of special ingredient (x) and feedrate (z) warns us that it will not be possible to distinguish between 'changes in impurity due to changes in x' and 'changes in impurity due to changes in z'.

Though the multiple regression equation containing x and t has been declared the 'best' equation on statistical grounds it may have occurred to you that there could be a multiple regression equation containing the feedrate (z) and other independent variables, which is almost as good on statistical grounds and more useful to the plant manager. We will now explore this possibility.

11.4   An alternative multiple regression equation

As we have already noted the computer program has a preference for weight of special ingredient (x) as the first independent variable. This selection is based on the correlation coefficients between the dependent variable (y) and the five independent variables (x, w, z, t and s) (Table 11.5).

Table 11.5   Correlation with percentage impurity

Variable	Weight of special ingredient	Catalyst age	Main ingredient		Agitation speed
			Feedrate	Inlet temperature	
	x	w	z	t	s
Correlation	-0.85	0.23	-0.78	0.52	0.11

So x is selected as the first independent variable and, as we have discovered, t is the second. The computer program stops at this point because the third independent variable (agitation speed) offers an increase in percentage fit which is not statistically significant. Had the program continued, however, we would have found that catalyst age (w) was the fourth variable and feedrate (z) was the last. With hindsight we can see that when x enters the regression equation z is excluded. This occurs, of course, because of the very high correlation (0.97) between x and z.

We can ensure that the weight of special ingredient (x) does *not* dominate the regression equations in this way if we instruct the computer program to fit an equation with impurity (y) as the dependent variable and with only *four* independent variables, z, w, t and s, from which to choose. When we adopt this strategy we find that the first variable to enter the equation is feedrate (z) and the second variable is inlet temperature (t) with the other two independent variables proving to be statistically non-significant. The fitted equations are:

$$y = 8.58 - 0.895z \qquad \text{60.8\% fit}$$

$$y = 5.44 - 0.94z + 1.66t \qquad \text{93.6\% fit}$$

We have already met the first of these two equations in the previous chapter and no further explanation is needed. The second equation is new to us and several points are worthy of note:

(a) The coefficient of z is -0.94 which implies that a unit increase in feedrate (z) would result in an impurity reduction of 0.94%. As we have stated earlier this figure must be treated with great suspicion because of the very high correlation between the two independent variables x and z.

(b) The coefficient of t is $+1.66$ which implies that a unit decrease in the inlet temperature would give an impurity reduction of 1.66%. This figure is in fairly close agreement with the 1.34% reduction offered by the 'best' equation ($y = 5.2 - 0.96x + 1.34t$) fitted earlier. In the light of these two equations therefore it would be reasonable for the plant manager to conclude that a reduction in impurity of approximately 1.5% could be obtained by reducing the inlet temperature by 1 unit.

(c) The percentage fit of the most recent equation (93.6%) is actually a little higher than the 93.4% achieved by the 'best' equation. It was pointed out earlier that the automatic variable selection methods used in some multiple regression packages don't always produce satisfactory results. It is possible that a different method of selection would have given a different 'best' equation. In order to get the best out of his computing facilities it is important for the scientist/technologist to 'interact' with the package, making full use of his accumulated experience.

We now find ourselves in the rather unsatisfactory position of having *two* multiple regression equations:

$$y = 5.2 - 0.96x + 1.34t$$

$$y = 5.44 - 0.94z + 1.66t$$

Though both equations have a high percentage fit we are unable to decide which of the two independent variables, x or z, is important. This uncertainty arises because of high correlation between x and z. In later chapters we will attempt to improve our position by extending the plant manager's experiment. Before doing so let us consider how the first of the two equations can be represented graphically.

11.5 Graphical representation of a multiple regression equation

In the previous chapter we had no difficulty in representing the simple regression equations in graphical form. We drew *two* axes at right angles; one axis for the dependent variable and one for the independent variable. To represent a regression equation which contains *two* independent variables we need *three* axes at right angles to each other. Such a system of axes requires a three-dimensional space and cannot be accommodated on a sheet of paper.

The multiple regression equation, $y = a + bx + ct$, could be represented in two dimensions, however, if we eliminated one of the independent variables.

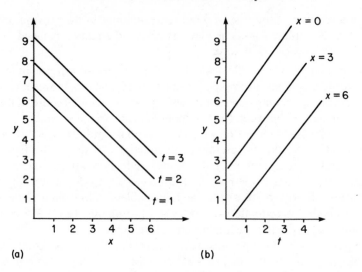

Figure 11.2   Two alternative representations of $y = 5.2 - 0.96x + 1.34t$

This can be achieved by giving that variable a fixed value. To reduce our equation $y = 5.2 - 0.96x + 1.34t$ we will fix the value of t at three suitable levels ($t = 1, t = 2$ and $t = 3$) to obtain the equations:

$$y = 6.5 - 0.96x \qquad \text{when } t = 1$$

$$y = 7.9 - 0.96x \qquad \text{when } t = 2$$

$$y = 9.2 - 0.96x \qquad \text{when } t = 3$$

These equations can be represented by the three straight lines in Fig. 11.2(a). Alternatively we can eliminate the other independent variable (x) from the equation by giving *it* suitable values. As the weight of special ingredient (x) varies from 0 to 6 within the ten batches it is reasonable to let x equal 0, 3 and 6. Doing so gives three simple equations which can be represented by the three straight lines in Fig. 11.2(b).

Either of the graphs, Fig. 11.2(a) or (b), might be of use to the plant manager in his search for operating conditions which will give a tolerable level of impurity. If he has concluded that weight of special ingredient (x) and inlet temperature (t) are the only two variables requiring attention then Fig. 11.2 may help him to compare the levels of impurity that are predicted for many different combinations of x and t.

You may have noticed that in Fig. 11.2(a) and in Fig. 11.2(b) we have a family of *parallel straight lines* and that, furthermore, the lines are equidistant. An equation of the form $y = a + bx + ct$ will always give a family of lines which are *parallel and straight*. Clearly such an equation is inadequate if you have a set of data which gives a graph resembling either of the scatter diagrams in Fig. 11.3.

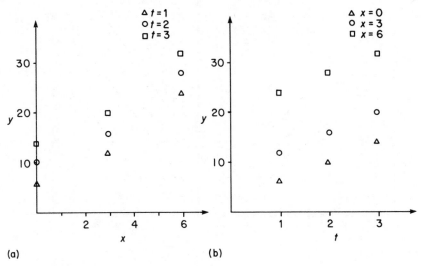

Figure 11.3   Set of data for which $y = a + bx + ct$ is unsuitable

11.6   Non-linear relationships

Fig. 11.3(a) and Fig. 11.3(b) are alternative representations of the same hypothetical set of data. Clearly the relationship between x and y is non-linear for each of the three values of t. Though the relationship between t and y *is* linear we can see in Fig. 11.3(b) that *equidistant* straight lines are inappropriate because of the curved relationship between x and y. The data which are plotted in Fig. 11.3 require an equation of the form:

$$y = a + bx + ct + dx^2$$

To fit such an equation we require *four* columns of data. Three of these columns will already be available because observations have been made of y, x and t and these values were used in plotting the graphs in Fig. 11.3. The values of x^2 could be calculated and then fed into the computer to give us the fourth column of data that we need. Fortunately this is rarely necessary because most multiple regression packages can generate *new variables* when asked to do so.

Returning to the data gathered by the plant manager in the previous chapter, we could ask the regression package to generate a quadratic term corresponding to each of the five independent variables originally measured. We would then have a total of ten independent variables (x, w, z, t, s, x^2, w^2, z^2, t^2 and s^2) which could be included in regression equations.

It is possible to generate variables having higher powers such as x^3 or z^4 but this is rarely done. In some applications logarithms, exponentials or trigonometric functions are found to be useful but will not be discussed here. Far more important is the ability to produce equations which will fit reasonably to data like those in Fig. 11.4

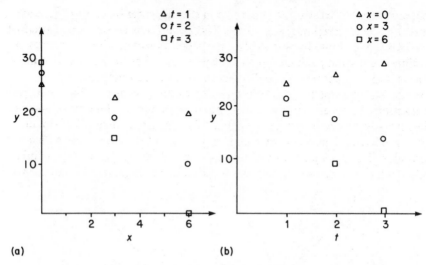

Figure 11.4   Set of data which illustrates an interaction between *x* and *t*

11.7   Interactions between independent variables

The data displayed in Fig. 11.4 do *not* exhibit curved relationships between the variables. In Fig. 11.4(a) we can see that a linear relationship exists between *x* and *y* for each value of *t*, but the straight lines are *not parallel*. Three straight lines are needed but each must have a different slope. Fig. 11.4(b) is an alternative representation of the data in Fig. 11.4(a) and again we see the need for non-parallel lines; for each value of *x* there exists a linear relationship between *t* and *y* but it is a different relationship in each case. The statistician would summarize Fig. 11.4 by saying that there appears to be an *interaction between x and t*.

> We say that there is an *interaction* between two independent
> variables when the effect of one (on the dependent vari-
> able) depends upon the value of the other.

The data represented in Fig. 11.4 suggest very strongly that the relationship between *x* and *y* depends on the value of *t* or, from the alternative standpoint, the relationship between *t* and *y* depends on the value of *x*. Interactions between pairs of independent variables are extremely important in research and development work. They will be discussed further in later chapters.

Our immediate concern is to find a multiple regression equation which will fit the data in Fig. 11.4. Such an equation must give non-parallel straight lines when plotted and this can be achieved by including a *cross-product term* (*dxt*), i.e.:

$$y = a + bx + ct + dxt$$

The new variable (xt) can be generated in the computer by multiplying each x value by the corresponding t value. The generation of such cross-product variables is even more important than the generation of quadratic variables in a multiple regression package for the simple reason that the detection of interactions is more important than the detection of curved relationships.

When we attempt to use a multiple regression package to find significant interactions between pairs of independent variables we find that there are *many* possibilities to investigate. With the data from ten batches of pigment we have only five independent variables $(x, w, z, t$ and $s)$ in the original data but these give rise to *ten pairs* of variables $(xw, xz, xt, xs, wz, wt, ws, zt, zs$ and $ts)$ each of which can be used as a cross-product term in a regression equation. Add

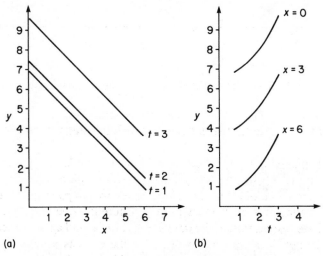

Figure 11.5 Two alternative representations of
$$y = 8.1 - 0.99x - 1.97t + 0.83t^2$$

to these the five quadratic terms $(x^2, w^2, z^2, t^2$ and $s^2)$ and we have a grand total of twenty variables which can be included in the right hand side of the regression equation. As we have only ten data points (i.e. ten batches) we are not in a good position to investigate all these possibilities, even if we ignore the crippling intercorrelation of x and z, but the following equations have been fitted to illustrate what we *might* be able to achieve when the plant manager's experiment is extended in a later chapter:

$$y = 8.1 - 0.99x - 1.97t + 0.83t^2 \qquad 96.7\% \text{ fit}$$
$$y = 3.5 - 0.37w + 0.321wt \qquad 46.7\% \text{ fit}$$

These two equations are represented in Figs 11.5 and 11.6. Fig. 11.5 portrays the quadratic equation whilst Fig. 11.6 portrays the equation containing the cross-product term (wt). The latter diagram illustrates an interaction between temperature (t) and catalyst age (w). We see that there is very little change in

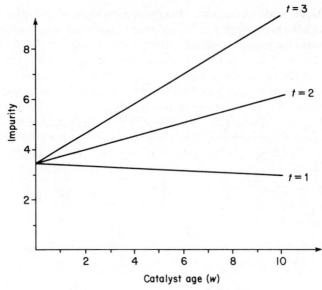

Figure 11.6   Graphical representation of

$$y = 3.5 - 0.37w + 0.321wt$$

impurity with age if a low temperature is used but a substantial increase in impurity as the catalyst ages if a high temperature is used.

11.8   Summary

In this chapter you have seen some of the power and versatility of multiple regression analysis as we have attempted to explain the batch to batch variation in pigment impurity. It is true that we have not been very successful but the following conclusions can now be drawn:

(a) The inlet temperature (t) of the main ingredient should be controlled in future batches and a reduction of one unit can be expected to give an impurity reduction of about 1.5%.

(b) The weight of special ingredient (x) and the feedrate (z) cannot be ignored, though we are unable to decide *which* of the two variables (or both) is important. We are unable to separate the effect of these variables because of the very bad design of the plant manager's experiment. The weakness of the experiment was revealed by inspecting the intercorrelations of the independent variables.

(c) Having found this deficiency in the experiment we must disregard the tentative conclusions drawn in Chapter 10.

(d) There is some indication that curved relationships might exist and that there might be an interaction between two of the independent variables. These will be explored later.

We must now set aside our techniques of statistical analysis whilst we consider the principles on which experiments should be based if we are to avoid some of the problems that we have encountered in this chapter.

Problems

(1) A research chemist is investigating the factors which determine the tensile strength (y) of a man-made fibre. He has carried out twelve production runs on small scale spinning and drawing plant using various values of the following independent variables:

(i) drying time of the polymer (w);
(ii) spinning temperature (x);
(iii) draw ratio (z).

Values of the response (y) and the three independent variables (w, x and z) are fed into a stepwise multiple regression package which gives the following regression equations:

$$y = 12.65 + 9.731z \qquad\qquad 50.7\% \text{ fit}$$

$$y = 53.81 + 9.731z - 0.1136x \qquad 69.3\% \text{ fit}$$

$$y = 68.73 + 5.824z - 0.0527x + 9.113w \qquad 73.8\% \text{ fit}$$

(a) Carry out t-tests to check the statistical significance of the three independent variables.
(b) By careful inspection of the equations and their percentage fits, what can you deduce about the correlations between the four variables?

(2) If you wish to use a computer-based multiple regression package or to discuss a multivariable problem with a statistician then your success will depend, at least in part, on your understanding of the terminology used in Chapters 10 and 11. You can test the extent to which you have absorbed this terminology by attempting to insert the missing words in the passage below. Don't be surprised if you need to refer back to the text or even to peep at the answers occasionally.

When using regression analysis we must distinguish between the dependent variable (which is often referred to as the (1)   and appears in the left hand side of the equation), on the one hand and the (2)   variables on the other. In a multiple regression equation we may have several (3)   variables but only one response. When choosing the independent variables to include in an equation we are not restricted to the variables that have been measured for we can also generate quadratic variables and (4)   variables. Inclusion of quadratic variables helps us to accommodate curved relationships whilst the use of cross product terms in the equation allows us to take account of (5)   between pairs of inde-

pendent variables. We say that there is an (6)   between two
independent variables when the effect of one variable upon the
(7)   depends upon the (8)   of the other. If, for
example we fitted the equation

$$y = a + bx + cxz + dz$$

and all three independent variables were found to be statistically significant
then we would conclude that there was an interaction between
(9)   and (10)   and it would be misleading to speak of
the effect of x on y without mentioning (11)

   To check the significance of each variable as it enters the equation we could
carry out a (12)   with the test statistic being calculated from the
new (13)   the old (14)   and $(n - k - 1)$, the latter
being known as the (15)   . If we put forward x, z and xz as independent variables and the fitted equations were

$$y = a + bx$$

then

$$y = c + dx + ez$$

and finally

$$y = f + gx + hz + ixz$$

we would know that:

(a) the correlation between y and (16)   would be greater than the
    correlation between y and z;
(b) at the second step the inclusion of z would offer a greater increase in
    (17)   than would inclusion of xz;
(c) the coefficient of x in the first equation (i.e. b) would only be equal to the
    coefficient of x in the second equation (i.e. d) if the correlation between x
    and z were equal to (18)   ;
(d) the coefficient of x in the *first* equation would have been calculated from
    the (19)   of x and y divided by the (20)   of x;
(e) the coefficient of x in the *second* equation would have been calculated by
    solving equations which included the covariance of (21)   and
    (22)   in addition to the covariance of x and y together with the
    covariance of z and y.

At each step in the stepwise regression procedure the equation is fitted by the
method of least squares. When using this method the values of the coefficients
are chosen so as to minimize the (23)   . As each new variable enters
the equation we get a further reduction in the residual sum of squares, unless
the correlation between:

 (i) the new independent variable and
(ii) the (24)   from the previous equation

is equal to zero, in which case the old and new equations will have the same residual sum of squares and the same (25)   . If we divide the residual sum of squares by its (26)   and then take the square root we obtain the (27)   which can be used in the calculation of confidence intervals for the true intercept, the true slopes and the true mean value of the (28)   variable for specified values of the independent variable(s). The residual standard deviation is a measure of the (29)   in the response variable that we would get with (30)   values of the independent variables that are included in the equation.

(3) A research and development chemist wishes to investigate the effect upon the tensile strength of a synthetic yarn, of varying the drying time of the polymer and the spinning temperature. Five runs of the spinning process under experimental conditions yield the following results:

Run	Drying time (min)	Spinning temperature (°C)	Tensile strength (kg)
A	60	240	3.5
B	70	290	3.9
C	30	210	3.1
D	70	260	3.7
E	45	250	3.8

These data are fed into a multiple regression computer package using the symbols:

x for drying time
z for spinning temperature
y for tensile strength.

The print-out from this package contains many regression equations but no information about goodness of fit or the statistical significance of the relationships.

Basic analysis

	x	z	y
Mean	55.0	250.0	3.60
SD	17.321	29.155	0.3162
C of V	31.5%	11.7%	8.8%

Correlation matrix

	x	z	y
x	1.000	0.8416	0.7303
z	0.8416	1.000	0.9220
y	0.7303	0.9220	1.000

Simple regression equations
(1) $y = 1.100 + 0.0100z$
(2) $z = -56.000 + 85.0000y$
(3) $y = 2.867 + 0.0133x$
(4) $x = -89.000 + 40.0000y$
(5) $x = -70.000 + 0.5000z$
(6) $z = 172.083 + 1.4167x$

Multiple regression equations
(7) $y = 0.9000 - 0.00286x + 0.01143z$
(8) $x = -51.6667 + 0.66667z - 16.6666y$
(9) $z = -1.9643 + 0.60714x + 60.7143y$

(a) Which of the six simple regression equations offers the best explanation of the variation in tensile strength from the yarn?
(b) Calculate the percentage fit for each of the six simple regression equations.
(c) Calculate the partial correlation coefficient for y and x with z as the fixed variable.
(d) Use the partial correlation from (c) to calculate the increase in percentage fit you would expect to get by introducing x into the equation $y = a + bz$.
(e) Calculate the percentage fit of the first multiple regression equation using the simple correlation coefficients from the matrix.
(f) Complete Tables 11.6 and 11.7 to obtain the residuals for the two simple regression equations which have z as the independent variable.

Table 11.6

Run	Actual y	Actual z	Predicted y $= 1.100 + 0.01z$	Residual
A	3.5	240	3.5	0.0
B	3.9	290	4.0	-0.1
C	3.1	210	3.2	-0.1
D	3.7	260	3.7	0.0
E				

Table 11.7

Run	Actual x	Actual z	Predicted x $= -70.0 + 0.5z$	Residual
A	60	240	50	10
B	70	290	75	-5
C	30	210	35	-5
D	70	260	60	10
E				

(g) Calculate the simple correlation between the residuals in Table 11.6 and the residuals in Table 11.7.

————————12————————
Some simple but effective experiments

12.1   Introduction

In Chapter 10 and Chapter 11 we made use of regression analysis in an attempt to draw conclusions from a set of data. Despite the great power and versatility of the statistical technique our efforts were not very successful because of intercorrelation between two of the independent variables. This correlation was a feature of the experiment and would not have arisen if the plant manager had used different values of feedrate (z) and/or weight of special ingredient (x) in some of the ten batches.

Having criticized the efforts of the plant manager we will now attempt the much more difficult task of giving him some positive advice. We will suggest what he *should* have done in order to obtain a set of data from which valid conclusions could be drawn. Armed with this advice he would be in a much stronger position should he decide to abandon his original experiment and to make a fresh start.

We can, however, suggest an alternative strategy which will surely appeal more strongly to the plant manager. This strategy will spell out precisely what he should *now* do in order to salvage as much as possible from the work he has carried out so far. He will need to manufacture some more batches, but when he has done so he will be able to draw valid conclusions concerning the effect of the five independent variables on the pigment impurity.

The ambitious programme we have just outlined cannot be implemented in this chapter, however, for we must *first* consider the relative advantages and disadvantages of certain simple experimental designs. We will, therefore, set aside the data that have held our attention so far and consider a variety of problems.

12.2   The classical experiment (one variable at a time)

An industrial chemist, employed by Trisell, wishes to investigate a process in which it suspected that the yield of a triacetate depends upon the *temperature* of the main ingredient at the start of the reaction and upon the *feedrate* of a secondary raw material. Past experience would suggest that temperatures between 55 °C and 70 °C might be suitable together with feedrates in the range 35 to 55. In order to find those values of temperature and feedrate which give maximum yield the researcher carries out an experiment in which he considers one variable at a time as follows:

Stage 1: To estimate the effect of changing temperature.

Using a feedrate of 40 units he carried out two trials using temperatures of 60°C and 70°C. The yields of triacetate obtained from the two trials are given in Table 12.1.

Table 12.1   Estimating the effect of temperature change

Temperature (°C)	Feedrate	Yield
60	40	76
70	40	72

The results of the two trials in stage 1 indicate that the lower temperature (60 °C) is giving the higher yield. We can make a quantitative estimate of the *effect* of temperature using:

$$\text{Temperature effect} = \text{yield at high temperature}$$
$$- \text{yield at low temperature}$$
$$= 72 - 76 = -4$$

This estimate suggests that the effect of increasing temperature (by 10 °C) is to *reduce* yield by 4 units.

Stage 2:   To estimate the effect of changing feedrate.

Since, in stage 1, the lower temperature (60 °C) gave the higher yield the researcher uses this temperature again in a third trial with a feedrate of 50. The yield from this trial is 70 which is included in Table 12.2 together with the yield from the earlier trial, which was carried out at the same temperature.

Table 12.2   Estimating the effect of feedrate change

Temperature (°C)	Feedrate	Yield
60	40	76
60	50	70

The results of the two trials in Table 12.2 indicate that the lower feedrate (40) is giving the higher yield and a quantiative estimate of the *effect* of feedrate is given by:

$$\text{Feedrate effect} = \text{yield at high feedrate}$$
$$- \text{yield at low feedrate}$$
$$= 70 - 76 = -6$$

This estimate suggests that the effect of increasing feedrate (by 10 units) is to *reduce* yield by 6 units.

From the two stages of this classical experiment the researcher draws the conclusion that future batches should be produced using a *starting temperature of 60 °C and a feedrate of 40*. His decision is supported by the simple graph of Fig. 12.1.

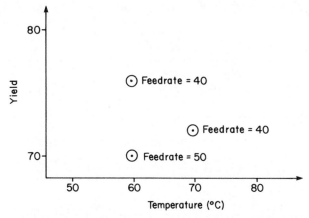

Figure 12.1   Data from Stages 1 and 2 of the experiment

One would hestitate, of course, to base an important decision on an experiment which consisted of only three trials, if it were possible to carry out a more extensive experiment. Putting aside this reservation let us question the usefulness of this *type* of experiment, in which one variable is changed at a time, by asking 'Is this experiment likely to lead the researcher to the operating conditions which give the maximum possible yield from his process?' Whilst it is unwise to give a simple yes or no answer to this question, it can be said most emphatically that this *classical approach to experimentation can lead the researcher badly astray*.

It is quite possible that the relationship between temperature, feedrate and yield for a chemical process is so complex that it would take many expensive experiments to unravel the true nature of this relationship. Suppose, for the sake of argument, that the true relationship is similar to that depicted in Fig. 12.2.

The two striking features of Fig. 12.2 are:

(a) For any value of feedrate the relationship between temperature and yield is represented by a *curve*, though the curves are 'fairly straight' in parts. To investigate a curved relationship a variable must have more than two levels.

(b) There is an *interaction* between temperature and feedrate.

The concept of an interaction was introduced in Chapter 11. Its importance cannot be emphasized too strongly since there must be few, if any, industrial

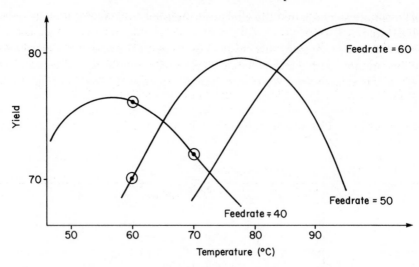

Figure 12.2   Hypothetical relationship between yield, temperature and feedrate

processes which do not have an interaction between two or more variables. A researcher who fails to recognize possible interactions when designing investigations or analysing data may well obtain results which are misleading.

12.3   Factorial experiments

How might we improve the classical experiment which is clearly inadequate in any situation where an interaction exists? First we will highlight the deficiency of the classical design by bringing together the results of Table 12.1 and Table 12.2 into a two-way table Table 12.3.

Table 12.3   Results of classical experiment

Feedrate	Temperature (°C)	
	60	70
40	76	72
50	70	

Since each variable has two levels, Table 12.3 has four cells. Three of the cells contain values of yield. These three cells represent the three trials carried out in the classical experiment. The empty cell corresponds to the treatment combination:

Feedrate = 50,   temperature = 70

If we were to carry out a fourth trial using this treatment combination then the four trials together would constitute an experiment which is known as a 2^2 *factorial experiment*. It is so called because we have two variables each at two levels and we have used *all* of the possible treatment combinations in the $2^2 = 4$ trials. (Note the notation. A 3^4 factorial experiment, for example, would have four factors, each at three levels, and would contain 81 trials.)

If we had carried out the fourth trial to complete the 2^2 factorial experiment the full set of results indicated by the relationship in Fig. 12.2 would be as in Table 12.4.

Table 12.4   Results of 2^2 factorial experiment

Feedrate	Temperature (°C)	
	60	70
40	76	72
50	70	78

What benefits do we get from this enlarged experiment which we did not get from the classical design? *One* benefit is that the interaction, which could not be seen in Table 12.3, stands out very clearly in Table 12.4. By comparing the two *rows* of the table we see that:

(a) increase in temperature causes an *increase* in yield, if we are using a feedrate of 50 units, whereas
(b) increase in temperature causes a *decrease* in yield, if we are using a feedrate of 40 units.

Alternatively, by comparing the two *columns* of Table 12.4 we see that:

(a) an increase in feedrate causes an *increase* in yield, if we are using a temperature of 70 °C, whilst
(b) an increase in feedrate causes a *decrease* in yield, if we are using a temperature of 60 °C.

Clearly the relationship between feedrate and yield depends upon the temperature used, or alternatively the relationship between temperature and yield depends upon the feedrate used. In a nutshell *there is an interaction between feedrate and temperature.* Perhaps the easiest way to spot an interaction between two factors is to compare the *diagonals* of a two-way table. In Table 12.4 the mean response for one diagonal is 77 whilst the mean response for the other diagonal is 71. Such a large difference (compared with some standard, which depends on residual variation) is an indication of an interaction.

A *second* benefit from carrying out the 2^2 factorial experiment rather than following the smaller classical design, is the increased accuracy with which we can estimate the effects of temperature and feedrate. We could of course use multiple regression to analyse the data from *any* experiment but with such simple designs the calculations can be carried out so easily that computing facilities are not needed. We will now consider how the effect estimates can be calculated.

12.4   Estimation of main effects and interaction (2² experiment)

We have already seen how the results of the classical experiment can be used to estimate the temperature effect and the feedrate effect. You may recall that we used

Temperature effect = yield at high level of temperature

− yield at low level of temperature

= 72 − 76 = −4

In the classical design we had three yield values but we were only able to use two of the three in the above calculation because the third yield value came from a trial in which a different level of feedrate (50) had been used. The classical design has an inherent lack of *balance* which is pointed out to us by the empty cell in Table 12.3. The 2² factorial design is balanced, in the sense that:

(a) we have two trials at each level of feedrate;
(b) we have two trials at each level of temperature.

To get the full benefit of the balanced design we change our method of calculation. At the same time we will refer to the effect of a variable as its *main effect* to distinguish it from interaction effects which we will also estimate.

> Main effect of a variable
>   = mean response at the high level of the variable
>
>   − mean response at the low level of the variable

This formula allows us to use *all four* response values in calculating each of the main effects.

Main effect of temperature = ½(72 + 78) − ½(76 + 70)

= 2

Main effect of feedrate = ½(70 + 78) − ½(76 + 72)

= 0

These calculations could have been performed very easily using the row means and the column means of Table 12.4. These are included in Table 12.5 together with the diagonal means which can be used in the calculation of an estimate of the interaction.

Table 12.5 Calculation of effect estimates

Feedrate	Temperature (°C)		Mean
	60	70	
40	76	72	74
50	70	78	74
Mean	73	75	74

– – 71

– – 77

It is clear from Table 12.5 that:

> Main effect of one variable = difference between row means
>
> Main effect of other variable = difference between column means

We can add to these formulae a third to be used for the estimation of the interaction effect.

> Interaction effect = difference between diagonal means
>
> = 77 – 71 = 6

The interaction effect could also be estimated using either:

Interaction = ½ (effect of temperature at high level of feedrate

 – effect of temperature at low level of feedrate)

or

Interaction = ½ (effect of feedrate at high level of temperature

 – effect of feedrate at low level of temperature)

All three formulae would give the same estimate of the interaction effect. Let us now bring together the results of our calculations which have given us three estimates:

Temperature effect = 2

Feedrate effect = 0

Temperature × feedrate interaction effect = 6

We can interpret these estimates as follows:

(a) The effect of increasing temperature (by 10 °C) is to increase the yield by 2 units, on average.

(b) The effect of increasing feedrate (by 10 units) is to cause no change in yield, on average.

(c) There is an interaction between feedrate and temperature. The existence of this interaction nullifies the value of statements (a) and (b) above. These statements are not incorrect, but they are worthless and potentially misleading. *Whenever an interaction exists it is unwise to speak of either of the two main effects in isolation.* The existence of the interaction tells us that the effect of temperature depends on the level of feedrate and that the effect of feedrate depends on the level of temperature. The relationship between the three variables, temperature, feedrate and yield is best described with reference to Table 12.5. This indicates that, for maximum yield, we should use *either* low values for both temperature and feedrate *or* high values for both independent variables.

12.5 Distinguishing between real and chance effects

From the results of the four trials in our 2^2 factorial experiment we have calculated three effect estimates. Had we carried out the four trials in a different order or on a different day, we would almost certainly have got different results and, consequently, different estimates. The estimated effects, therefore, are unlikely to be exactly equal to the true effects and we cannot dismiss the possibility that the true effects could be equal to zero. How are we to decide whether the independent variables really do influence the yield or not?

The reader will not be surprised to learn that we can use a significance test to distinguish between real and chance effects. The basis of this test will be clearer if we introduce a model to illustrate the relationship between the independent variables and the response. You will recall that, in earlier chapters, we used models such as

$$y = \alpha + \beta x + \text{error}$$

and

$$y = \alpha + \beta x + \gamma t + \text{error}$$

Such models were appropriate because we were fitting regression equations. For the method of analysis that we have used with the results of a 2^2 factorial experiment a rather different model is appropriate as in Table 12.6.

In this model y_1, y_2, y_3 and y_4 represent the observed yields whist μ represents the true yield at the *centre* of the experimental region, i.e. with temperature equal to 65 and feedrate equal to 45. T represents the true temperature main effect, F represents the true feedrate main effect and I represents the true interaction effect. (It would have been appropriate to use Greek letters rather than T, F and I but the author is reluctant to introduce three more Greek letters at this point.) Each observed yield will differ from the true yield corresponding to the temperature and feedrate used in the manu-

Table 12.6 2^2 factorial experiments: a model

	Temperature ($^{\circ}C$)	
Feedrate	60	70
40	$y_1 = \mu - \dfrac{T}{2} - \dfrac{F}{2} + \dfrac{I}{2} + e_1$	$y_2 = \mu + \dfrac{T}{2} - \dfrac{F}{2} - \dfrac{I}{2} + e_2$
50	$y_3 = \mu - \dfrac{T}{2} + \dfrac{F}{2} - \dfrac{I}{2} + e_3$	$y_4 = \mu + \dfrac{T}{2} + \dfrac{F}{2} + \dfrac{I}{2} + e_4$

facture of the batch. The differences between observed and true yields are represented by the four errors (e_1, e_2, e_3 and e_4) in the model.

We have already calculated estimates for the two main effects and the interaction (i.e. we have estimated that $T = 2$, $F = 0$ and $I = 6$). Using the mean yield of all four cells we can estimate that μ is equal to 74. Having estimated T, F, I and μ, however, we can make no further progress. It is not possible, for example, to estimate the four errors e_1, e_2, e_3 and e_4. You could argue that the four errors are of little or no interest because each error is *random* and, therefore, never likely to be repeated. It is true that the individual errors are not important in themselves, but it is also true that a knowledge of the whole population of errors would be useful. If we knew, for example, that the errors came from a population which had a normal distribution with a mean of zero and a standard deviation of 2.5, say, then we would be able to decide whether the effect estimates were likely to have arisen simply by chance.

Unfortunately we cannot obtain an estimate of the error standard deviation from our 2^2 factorial experiment. To calculate such an estimate we would need the results of at least two trials carried out under *identical* conditions. As each of the four batches in our experiment was manufactured with a unique combination of temperature and feedrate we do not have a means of estimating the error standard deviation, or residual standard deviation as it is often called.

> A 2^n factorial experiment is a very efficient means of estimating main effects and interactions but it does not give an estimate of residual variation.

There are several ways in which we could obtain an estimate of the residual standard deviation. One possibility is to repeat the whole of the 2^2 factorial experiment. The yield values from the four new batches could be combined with the yield values of the four original batches; then, from the enlarged set of data, we would be able to estimate the two main effects (F and T) and the interaction (I) *plus* the residual standard deviation. (It would be better to mingle the two factorial experiments and carry out the eight trials in random order, as we shall see later.) The benefits to be gained by carrying out *two replicates* of a 2^2 factorial experiment may be clearer if we examine the model in Table 12.7.

Table 12.7   Two replicates of 2^2 factorial experiment: a model

Feedrate	Temperature (°C)	
	60	70
40	$y_1 = \mu - \dfrac{T}{2} - \dfrac{F}{2} + \dfrac{I}{2} + e_1$	$y_2 = \mu + \dfrac{T}{2} - \dfrac{F}{2} - \dfrac{I}{2} + e_2$
40	$y_5 = \mu - \dfrac{T}{2} - \dfrac{F}{2} + \dfrac{I}{2} + e_5$	$y_6 = \mu + \dfrac{T}{2} - \dfrac{F}{2} - \dfrac{I}{2} + e_6$
50	$y_3 = \mu - \dfrac{T}{2} + \dfrac{F}{2} - \dfrac{I}{2} + e_3$	$y_4 = \mu + \dfrac{T}{2} + \dfrac{F}{2} + \dfrac{I}{2} + e_4$
50	$y_7 = \mu - \dfrac{T}{2} + \dfrac{F}{2} - \dfrac{I}{2} + e_7$	$y_8 = \mu + \dfrac{T}{2} + \dfrac{F}{2} + \dfrac{I}{2} + e_8$

It should be clear from Table 12.7 that if the two yield values in any particular cell differ from each other, then this difference can only be due to random error and is not caused by changes of temperature or feedrate. We could, therefore, use the differences within the four cells to calculate the residual standard deviation. An alternative approach, which is more consistent with procedures we have used earlier, starts with the calculation of a mean and standard deviation for each of the four cells in Table 12.7. The cell means are then used to calculate the effect estimates whilst the cell standard deviations are combined to give an estimate of the residual standard deviation. You may recall that we combined several standard deviations in Chapter 7 using:

$$\text{Combined standard deviation} = \sqrt{\left\{ \frac{\Sigma\,[(\text{d.f.})\,(\text{SD}^2)]}{\Sigma\,(\text{d.f.})} \right\}}$$

As each of our four standard deviations has 1 degree of freedom this equation becomes:

$$\text{Residual standard deviation (RSD)} = \sqrt{\frac{s_1^2 + s_2^2 + s_3^2 + s_4^2}{4}}$$

To illustrate this approach let us assume that we have carried out two replicates of the 2^2 factorial experiment and the results of the eight trials are the yield values in Table 12.8.

The four cell standard deviations in Table 12.8 (see Table 12.9) can be combined as follows:

Residual standard deviation (RSD) =

$$\sqrt{\left[\frac{(1.414)^2 + (1.414)^2 + (4.243)^2 + (2.828)^2}{4} \right]}$$

$$= 2.739$$

Table 12.8   Yields of eight batches in a 2×2^2 factorial experiment

Feedrate	Temperature (°C)	
	60	70
40	$y_1 = 77$ $y_5 = 75$	$y_2 = 73$ $y_6 = 71$
50	$y_3 = 67$ $y_7 = 73$	$y_4 = 80$ $y_8 = 76$

Table 12.9   Cell means and standard deviation for the data in Table 12.8

Feedrate	Temperature (°C)	
	60	70
40	$\bar{y}_1 = 76.0$ $s_1 = 1.414$	$\bar{y}_2 = 72.0$ $s_2 = 1.414$
50	$\bar{y}_3 = 70.0$ $s_3 = 4.243$	$\bar{y}_4 = 78.0$ $s_4 = 2.828$

If we were to manufacture a series of batches under identical conditions then we would expect the standard deviation of the batch yields to be approximately 2.739. This is not a very good estimate of batch to batch variability, of course, as it is based on only four degrees of freedom, but it is better than no estimate at all.

The four cell means in Table 12.8 are exactly equal to the four yield values in Table 12.5. This larger experiment will, therefore, give *exactly the same* effect estimates that we obtained from the original 2^2 factorial experiment (i.e. $T = 2$, $F = 0$ and $I = 6$). It is important to note, however, that we can place more confidence in these new estimates because each is based on eight yield values, whereas each of the original effect estimates was based on only four. This point is taken into account when we carry out a *t*-test to check the statistical significance of the effect estimates. The test statistic for the *t*-test is calculated as follows:

(Effect estimate)/RSD   for a 2^2 factorial experiment

(Effect estimate)/(RSD/$\sqrt{2}$)   for two replicates of a 2^2 factorial experiment.

> When carrying out a *t*-test on the effect estimates obtained from p replicates of a 2^n factorial experiment the test statistic can be calculated from:
>
> $$(\text{Effect estimate})/[\text{RSD}/\sqrt{(p\,2^{n-2})}]$$
>
> where RSD is the residual standard deviation.

For a discussion of the above formula the reader should refer to Appendix F. We will make use of the formula to carry out t-tests on the effect estimates and we will start with the largest of the three.

Null hypothesis – There is no interaction between temperature and feedrate (i.e. $I = 0$).

Alternative hypothesis – There is an interaction between temperature and feedrate (i.e. $I \neq 0$).

$$\text{Test statistic} = \frac{\text{effect estimate}}{\text{RSD}/\sqrt{2}}$$

$$= \frac{6}{2.739/\sqrt{2}}$$

$$= 3.10$$

Critical values – From the two-sided t-table with 4 degrees of freedom:

   2.78 at the 5% significance level
   4.60 at the 1% significance level.

Decision – We reject the null hypothesis at the 5% level of significance.

Conclusion – We conclude that there is an interaction between temperature and feedrate.

Carrying out similar t-tests on the two main effect esimates would give inconclusive results in that we would be unable to reject the null hypothesis in either case. If, however, we had a better estimate of the residual standard deviation we might have found the temperature main effect to be significant also. A better estimate of the residual standard deviation could be obtained by either:

(a) carrying out further replicates of the 2^2 factorial experiment;
(b) manufacturing several batches of triacetate under fixed conditions (n batches would give an estimate of RSD with $(n-1)$ degrees of freedom);
(c) referring to the plant records to extract the yields of several batches which were produced during a period when conditions were stable. From these yields an estimate of the residual standard deviation could be calculated.

You might imagine that the third alternative would be dangerous. If we obtained an estimate of residual standard deviation that was much smaller than the true value then there would be an increased risk of declaring non-existent effects to be significant. In reality the reverse is more likely to occur. An estimate of residual standard deviation obtained from routine plant data is likely to be *inflated* because of unintended variation in operating conditions. Thus we would incur a greater risk of failing to detect real effects by using such an estimate.

Before we progress to consider a factorial experiment with three independent variables we will return to the results of the 2^2 factorial experiment for a second analysis. This time we will use multiple regression analysis.

12.6   The use of multiple regression with a 2^2 experiment

The results of the 2^2 factorial experiment are set out in Table 12.4. This style of tabulation proved to be very useful when we needed to calculate the row, column and diagonal means but it is not the type of table that we used in earlier chapters. Before we carry out a regression analysis we will set out the data in the manner (Table 12.10) that would be expected by a multiple regression package.

Table 12.10   Data from the 2^2 factorial experiment

	Temperature x	Feedrate z	Yield y
	60	40	76
	70	40	72
	60	50	70
	70	50	78
Mean	65.0	45.0	74.0
SD	5.77	5.77	3.65

If we wish to analyse the data in Table 12.10 by means of our multiple regression package we must specify that y is the dependent variable whilst x and z are independent variables. In addition we will specify that the program should add the independent variables, one by one, to the equation *without* checking the statistical significance. In this way we can be sure that the final equation will contain *all* of the independent variables, including those which are significant and those which are not. The program responds by printing out the following equations:-

$$y = 61 + 0.2x \qquad 10\% \text{ fit}$$
$$y = 61 + 0.2x + 0.0z \qquad 10\% \text{ fit}$$

At first sight these equations may appear to have no connection with the analysis carried out earlier on these same data. You will recall that we calculated the three effect estimates to be:

$$\text{Temperature effect} = 2$$
$$\text{Feedrate effect} = 0$$
$$\text{Temperature} \times \text{feedrate interaction} = 6$$

On further inspection of the two alternative analyses we find a certain measure of agreement between them, the following points being worthy of note:

(a) The coefficient of x (i.e. $+0.2$) in the regression equations indicates a yield increase of 0.2 for a unit increase in temperature. This is in perfect agreement with our earlier estimate of $+2$ for the temperature effect since this latter figure relates to a total temperature change of 10 degrees.
(b) The coefficient of z (i.e. 0.0) in the regression equations is in agreement with our earlier estimate of the feedrate effect (i.e. 0).
(c) The regression analysis has not embraced the temperature $\times$ feedrate interaction, but this can be rectified by introducing a cross-product term.

Perhaps our strongest memory of the previous chapter is the trouble that was caused by the intercorrelation of two independent variables. In fact the main reason for considering experimental designs was to obtain data in which such things did not arise.

To see how the 2^2 factorial design matches up to our requirements we can examine the correlation matrix printed out by the regression package in Table 12.11.

Table 12.11   Correlation matrix for 2^2 experiment

	x	z	y
x	1.00	0.00	0.316
z		1.00	0.00
y			1.00

In Table 12.11 the dependent variable (y) is separated from the independent variables (x and z) by the dashed lines. We see that the correlation between the independent variables, x and z, is equal to zero. This is highly satisfactory.

Let us now extend our regression equation to include a cross-product term in order to assess any interaction between temperature and feedrate. To do this we must generate a new variable so that our complete set of data is that in Table 12.12.

If the data of Table 12.12 are fed into the multiple regression package they yield the following equations:

$$y = 61 + 0.2x \qquad\qquad\qquad\qquad 10\% \text{ fit}$$

$$y = 61 + 0.168235x + 0.000706xz \qquad 10.53\% \text{ fit}$$

$$y = 412 - 5.12x + 0.12xz - 7.8z \qquad 100\% \text{ fit}$$

The first equation is already familiar to us. It is identical with the first equation printed out earlier and it has the same percentage fit (10%).

You may have expected, however, that the cross-product variable (xz)

Table 12.12   Extended data for the 2^2 experiment

	Temperature	Feedrate	Temperature × feedrate	Yield
	x	z	xz	y
	60	40	2400	76
	70	40	2800	72
	60	50	3000	70
	70	50	3500	78
Mean	65.0	45.0	2925.0	74.0
SD	5.77	5.77	457.35	3.65

would have been the first to enter the equation. After all, it was the *interaction* between feedrate and temperature which was earlier found to be significant whilst the two main effects were *not*.

When the cross-product term *does* enter the regression equation the percentage fit increases from 10% to 10.53%. This minute increase is in stark contrast to the massive jump which occurs when the feedrate (z) finally enters the equation. When you recall that the feedrate effect was earlier estimated to be *zero* you will realize that multiple regression analysis has again revealed its darker side.

Once again the trouble was caused by *intercorrelation of the independent* variables. This is revealed if we examine the correlation matrix printed out by the regression package and reproduced in Table 12.13. Though there is no correlation between the two original variables, x and z, each one of these *is* correlated with the newly introduced variable (xz).

Table 12.13   Correlation matrix

	x	z	xz	y
x	1.00	0.00	0.568	0.316
z		1.00	0.821	0.00
xz			1.00	0.240
y				1.00

The two offending correlations are 0.568 and 0.821. Neither of these is as large as the 0.97 that we found between two independent variables when analysing a different set of data in the previous chapter. Nonetheless their combined effect is to give a series of equations which could be misleading and they serve to remind us yet again that the use of multiple regression analysis can be dangerous.

Fortunately, with data from a 2^2 factorial experiment the intercorrelation of the independent variables can be *eliminated* before equations are fitted. This is achieved by *standardizing* (or scaling or transforming) the values of the two independent variables x and z as follows:

$$X = (x - \bar{x})/(\text{half range of } x)$$

$$Z = (z - \bar{z})/(\text{half range of } z)$$

i.e.

$$X = (x - 65.0)/5.00$$

$$Z = (z - 45.0)/5.0$$

After carrying out this transformation the data from the 2^2 factorial experiment are as given in Table 12.14.

Table 12.14   Transformed data

	X	Z	XZ	y
	−1	−1	+1	76
	+1	−1	−1	72
	−1	+1	−1	70
	+1	+1	+1	78
Mean	0.00	0.00	0.00	74.00
SD	1.00	1.00	1.00	3.65

Submitting this transformed data set to the multiple regression package results in the following equations being printed out:

$$y = 74.0 + 3.0XZ \qquad \text{90\% fit}$$

$$y = 74.0 + 3.0XZ + 1.0X \qquad \text{100\% fit}$$

$$y = 74.0 + 3.0XZ + 1.0X + 0.0Z \qquad \text{100\% fit}$$

This set of equations is much more reasonable. The first variable to enter the equation is XZ which represents the interaction effect. The percentage fit of this first equation 90%. The second independent variable to enter the equation is X which adds a further 10% to the percentage fit whilst the third variable Z adds nothing.

The coefficients of the three variables in the third equation (3.0, 1.0, 0.0) are also in agreement with the effect estimates calculated earlier (6.0, 2.0, 0.0) if a factor of 2 is taken into account. This 2 arises because X, Z and XZ are varying from −1 to +1 which is a total change of 2 units.

By the use of variable transformation (or *scaling* as it is often called) we have managed to obtain regression equations which would lead us to conclusions rather similar to those suggested by the effect estimates calculated from the two-way table (Table 12.5). Without the use of scaling the multiple regression package gives us a sequence of equations which could lead someone to very different conclusions. It is possible, for example, that an inexperienced data analyst might conclude that the *least* important of the independent variables (i.e. z) was actually the *most* important.

In fairness to the statistical technique and in summary of this discussion the following points should be noted:

(a) The use of multiple regression analysis to analyse the data from a 2^2 factorial experiment is neither necessary nor desirable. As we have no way of feeding into the program an estimate of residual variance, no sensible t-tests can be carried out.

(b) As the 2^2 factorial experiment gives only four data points we are certain to get 100% fit when we include the three independent variables (x, z and xz) in the equation.

(c) By scaling (or transforming) the independent variables we get a much more meaningful set of equations. The need for scaling arises when we introduce the cross-product variable (xz) which is correlated with x and with z unless scaling is used..

(d) The use of scaling would *not* have eliminated the intercorrelation that was so troublesome in the previous chapter. That particular correlation was between two of the *measured* variables. It resulted from bad design and not from the introduction of extra variables.

A warning

Whenever cross-product variables and/or quadratic variables are introduced into a regression equation the independent variables should be scaled.

12.7   A 2^3 factorial experiment

A research and development chemist is attempting to increase the tensile strength of a particular type of rubber so that it will be suitable for a new application. He realizes that several investigations may be required before a worthwhile increase in tensile strength is achieved and he limits the first investigation to examining the effects of certain changes in the formulation. He has selected three variables for inclusion in a 2^3 factorial experiment:

(a) the carbon black content;
(b) the type of accelerator;
(c) the percentage of natural rubber in the polymer.

A list of the eight trials is drawn up and random numbers are used to put the trials into random order. This is the order in which the trials are carried out and the results are listed in Table 12.15. The use of randomization in the execution of experiments is a means of avoiding unsuspected systematic errors. For a full discussion of randomization see Cox (1958).

Table 12.15   Results of a 2^3 factorial experiment

Carbon black (parts per hundred) x	Type of accelerator z	% natural rubber in polymer w	Tensile strength (kg/cm²) y
20	HLX	50	236
40	HLX	75	284
40	MD	75	272
20	MD	75	232
40	MD	50	280
40	HLX	50	264
20	HLX	75	256
20	MD	50	248

The chemist does not expect to draw final conclusions from this experiment. If he had had such expectations he would have used a different design which will be discussed in a later chapter. He simply hopes that an analysis of the variation in tensile strength in Table 12.15 will enable him to decide what further trials need to be carried out. Note that in Table 12.15 the variables have been labelled x, z and w. In addition to renaming the variables we will also refer to the high level and the low level of each variable as in Table 12.16.

Table 12.16   Levels of the independent variables

Variable	Low level	High level
x (carbon black)	20	40
z (type of accelerator)	MD	HLX
w (% natural rubber)	50	75

To facilitate the analysis of the results of this experiment we subtract 200 from each response value in Table 12.15 and transfer the reduced figure to a standard three-way table, Table 12.17.

Table 12.17   Three-way table of response values

	z			
	Low		High	
	w		w	
x	Low	High	Low	High
Low	48	32	36	56
High	80	72	64	84

From Table 12.17 we could calculate estimates of the three main effects using the formula that we used with a 2^2 design:

Main effect of a variable

= mean response at the high level of the variable

− mean response at the low level of the variable

It would not, however, be obvious how we could calculate estimates of the interaction effects from this table, so we will first break down the three-way table into three two-way tables. This is achieved by averaging the response values in pairs to give Table 12.18.

Table 12.18   Two-way tables of mean response values

(a)			(b)			(c)		
	z			w			w	
x	Low	High	x	Low	High	z	Low	High
Low	40	46	Low	42	44	Low	64	52
High	76	74	High	72	78	High	50	70

Some care is needed in completing the two-way tables. The entry of 40 in Table 12.18(a) is the mean of 48 and 32 from Table 12.17, whilst the entry 52 in Table 12.18(c) is the mean of 32 and 72 from Table 12.17. From each two-way table we can estimate two main effects and a two factor interaction using the row means, column means and diagonal means as follows:

From Table 12.18(a) we can calculate estimates of:

Main effect x $= \frac{1}{2}(74 + 76) - \frac{1}{2}(46 + 40) = 32$

Main effect z $= \frac{1}{2}(74 + 46) - \frac{1}{2}(76 + 40) = 2$

Interaction xz $= \frac{1}{2}(74 + 40) - \frac{1}{2}(76 + 46) = -4$

From Table 12.18(b) we can calculate estimates of:

Main effect x $= \frac{1}{2}(78 + 72) - \frac{1}{2}(44 + 42) = 32$

Main effect w $= \frac{1}{2}(78 + 44) - \frac{1}{2}(72 + 42) = 4$

Interaction xw $= \frac{1}{2}(78 + 42) - \frac{1}{2}(72 + 44) = 2$

From Table 12.18(c) we can calculate estimates of:

Main effect z $= \frac{1}{2}(50 + 70) - \frac{1}{2}(52 + 64) = 2$

Main effect w $= \frac{1}{2}(70 + 52) - \frac{1}{2}(64 + 50) = 4$

Interaction zw $= \frac{1}{2}(70 + 64) - \frac{1}{2}(50 + 52) = 16$

Each of the three main effect estimates can be calculated from either of two tables. The estimate is of course the same whichever table is used. Since our 2^3 experiment contains three variables we can also estimate the *three-variable interaction* using the following calculation which is based on Table 12.17:

$$
\begin{aligned}
\text{Interaction } xzw \ &= \ \tfrac{1}{2}\{\text{interaction } xw \text{ with } z \text{ at high level} \\
&\quad - \text{interaction } xw \text{ with } z \text{ at low level}\} \\
&= \ \tfrac{1}{2}\{[\tfrac{1}{2}(84+36)-\tfrac{1}{2}(64+56)] \\
&\quad -[\tfrac{1}{2}(48+72)-\tfrac{1}{2}(80+32)]\} \\
&= \ -2
\end{aligned}
$$

In many situations it is difficult to attach any meaning to a three-variable interaction. The interaction xzw tells us 'how the interaction xw depends upon the level of z', or alternatively 'how the interaction xz depends upon the level of w', or alternatively 'how the interaction zw depends on the level of x'. Taking the first of these three statements and translating it into the language of our experiment the three-factor interaction tells us how the interaction between the carbon black content and the percentage of natural rubber depends upon the type of accelerator used. Whether or not it is *possible* for such an interaction to exist can only be decided on chemical/physical grounds. We will return to this point later.

Drawing together the results of our calculations we have the seven effect estimates listed below:

Main effect x	$= 32$
Main effect z	$= 2$
Main effect w	$= 4$
Interaction xz	$= -4$
Interaction xw	$= 2$
Interaction zw	$= 16$
Interaction xzw	$= -2$

These numerical estimates will be more meaningful if we return to the language of the chemist, from the language of the statistician in which the list is written. The value of 32 for our estimate of main effect x implies that the effect of increasing carbon black from 20 parts/hundred to 40 parts/hundred is to increase the tensile strength by 32 kg/cm^2. The second entry in the list implies that the effect of changing the accelerator type from MD to HLX increases the tensile strength by 2 kg/cm^2.

Interpreting a two-factor interaction effect is best achieved by reference to a two-way table. The nature of interaction zw, for which we have such a large effect estimate, can be illustrated by Table 12.18(c) which tells us that a high tensile strength will result from using the HLX accelerator and 75% natural rubber in the polymer blend, whilst the second best combination of these two factors is to use the MD accelerator with 50% natural rubber. Clearly the effect of increasing % natural rubber (w) is to increase the tensile strength when we use the HLX accelerator but to decrease tensile strength if we use the MD accelerator.

Having calculated the effect estimates it would be wise to check their statistical significance before acting upon the conclusions they appear to imply. Unfortunately we cannot carry out a t-test upon the estimates without a suitable standard deviation. We have already noted that a 2^n factorial experiment does not yield such an estimate. Let us see what we can achieve with multiple regression analysis.

12.8   The use of multiple regression with a 2^3 experiment

Before we explore various methods of obtaining an estimate of residual variance we will make use of our multiple regression package to analyse the data from the 2^3 factorial experiment. A prerequisite for this course of action is to quantify the qualitative variable, 'type of accelerator (z)'. If we replace MD with -1 and replace HLX with $+1$ there will be no necessity for scaling of this variable. In the light of our earlier experience we will, of course, scale the other two measured variables using:

$$X = (x - 30.0)/10.00 \quad \text{and} \quad W = (w - 62.5)/12.50$$

Introduction of cross-product variables gives us the complete set of data in Table 12.19.

Table 12.19 2^3 experiment – data scaled for regression analysis

	X	Z	W	XZ	XW	ZW	XZW	y
	-1	$+1$	-1	-1	$+1$	-1	$+1$	236
	$+1$	$+1$	$+1$	$+1$	$+1$	$+1$	$+1$	284
	$+1$	-1	$+1$	-1	$+1$	-1	-1	272
	-1	-1	$+1$	$+1$	-1	-1	$+1$	232
	$+1$	-1	-1	-1	-1	$+1$	$+1$	280
	$+1$	$+1$	-1	$+1$	-1	-1	-1	264
	-1	$+1$	$+1$	-1	-1	$+1$	-1	256
	-1	-1	-1	$+1$	$+1$	$+1$	-1	248
Mean	0.0	0.0	0.0	0.0	0.0	0.0	0.0	259.0
SD	1.00	1.00	1.00	1.00	1.00	1.00	1.00	18.19

If we now specify that the dependent variable is y and the independent variables are X, Z, W, XZ, XW, ZW and XZW the regression package prints out the correlation matrix (Table 12.20) and the following equations:

Table 12.20   Correlation matrix for 2^3 experiment

	X	Z	W	XZ	XW	ZW	XZW	y
X	1.00	0.00	0.00	0.00	0.00	0.00	0.00	0.879
Z		1.00	0.00	0.00	0.00	0.00	0.00	0.055
W			1.00	0.00	0.00	0.00	0.00	0.100
XZ				1.00	0.00	0.00	0.00	−0.110
XW					1.00	0.00	0.00	0.055
ZW						1.00	0.00	0.440
XZW							1.00	−0.055
y								1.00

$$y = 259.0 + 16.0X \qquad 77.34\% \text{ fit}$$

$$y = 259.0 + 16.0X + 8.0ZW \qquad 96.68\% \text{ fit}$$

$$y = 259.0 + 16.0X + 8.0ZW - 2.0XZ \qquad 97.89\% \text{ fit}$$

$$y = 259.0 + 16.0X + 8.0ZW - 2.0XZ + 2.0W \qquad 99.09\% \text{ fit}$$

$$y = 259.0 + 16.0X + 8.0ZW - 2.0XZ + 2.0W$$
$$- 1.0XZW \qquad 99.40\% \text{ fit}$$

$$y = 259.0 + 16.0X + 8.0ZW - 2.0XZ + 2.0W$$
$$- 1.0XZW + 1.0XW \qquad 99.70\% \text{ fit}$$

$$y = 259.0 + 16.0X + 8.0ZW - 2.0XZ + 2.0W$$
$$- 1.0XZW + 1.0XW + 1.0Z \qquad 100.00\% \text{ fit}$$

There is substantial agreement between the regression analysis above and the effect estimates calculated earlier. The following points should be noted:

(a) The order in which the independent variables enter the regression equation corresponds exactly with the size of the effect estimates. For example, the largest estimate (32) indicates that carbon black (x) is the most important variable and we see that X is the first variable to enter the equation.
(b) The coefficients of the variables already in the regression equation do not change as new variables enter. This is to be expected when the intercorrelations amongst the independent variables are all equal to zero as we see in Table 12.20.

Do not forget that the automatic significance testing was suppressed whilst the regression package produced the succession of equations which culminated in 100% fit for the seventh equation. If we were to carry out the significance tests we would find that the first two independent variables, X and ZW, were significant but the third was not. This would lead us to the conclusion that the 'carbon black content (x)' and the 'interaction between type of accelerator and % natural rubber' were important. The significance tests are not based on a 'good' estimate of the residual variance because, as we noted earlier, no two trials were given the same treatment combination. Accepting the results of the significance tests is tantamount to assuming that the other effects (z, w, xz, xw and xzw) do not exist. Scientists and technologists are very reluctant to make such assumptions when analysing data by hand but they are often willing to let a computer program make such assumptions on their behalf.

We have returned to the point made earlier that a *2^n factorial experiment does not give us a reliable estimate of residual variance*. In practice there are three ways in which this problem can be overcome:

(a) An estimate of residual variance may have been available *before* the experiment was carried out. We will assume that the research and development chemist carrying out this experiment had no such estimate.

(b) Further trials can be carried out using one or more of the eight treatment combinations or even using a new treatment combination that was not included in the original experiment. One obvious possibility is to repeat (or replicate) the whole of the 2^3 factorial experiment and we will investigate the usefulness of this strategy in the next section.

(c) An *assumption* can be made that one or more of the interactions could not possibly exist. If, for example we assume that the three factor interaction (i.e. interaction xzw) does not exist then it would be possible to calculate an estimate of the residual standard deviation that had 1 degree of freedom. Two points need to be stressed immediately:

   (i) An estimate of residual variance which is based on only 1 degree of freedom is of doubtful value (the critical value from the t-table using 1 degree of freedom is 12.71 at 5% significance).

   (ii) Any assumption concerning the non-existence of an interaction is clearly very dangerous and should certainly be based on chemical rather than statistical reasoning. Note, however, that such an assumption is conservative in the sense that, if the assumption is invalid, the residual will be inflated and we may fail to detect an effect which actually exists.

12.9   Two replicates of a 2^3 factorial experiment

The research and development chemist decides to replicate the 2^3 factorial experiment. He takes great care to use the same levels of the variables which were used in the first experiment and he goes to considerable lengths to ensure that extraneous variables which were controlled in the first experiment are similarly controlled throughout the eight trials of this second experiment. He ensures, for example, that the natural rubber and the carbon black come from the same batch that was used previously and that the milling of the rubber is carried out at the same speed and for the same duration. The order of the eight trials is randomized and the response values are included with those from the first experiment in Table 12.21.

Statistics in Research and Development

Table 12.21   Two replicates of a 2^3 factorial experiment

			z		
		Low		*High*	
		w		*w*	
x		*Low*	*High*	*Low*	*High*
Low		48	32	36	56
		52	32	32	48
High		80	72	64	84
		76	80	56	80

The two response values in any cell of Table 12.21 are not simply repeat determinations of tensile strength on the same batch but are single determinations on two batches produced with the same formulation. The variability within any cell is *not* caused by changes in the independent variables but is a manifestation of unassignable variation or error. The first step in our analysis of the above data is to compute a table of cell means and cell standard deviations, as in Table 12.22.

Table 12.22   Means and standard deviations of data in Table 12.21

				z	
		Low		*High*	
		w		*w*	
x		*Low*	*High*	*Low*	*High*
Low		50.0	32.0	34.0	52.0
		2.828	0.000	2.828	5.657
High		78.0	76.0	60.0	82.0
		2.828	5.657	5.657	2.828

The cell standard deviations can now be combined to obtain an estimate of the residual standard deviation with 8 degrees of freedom.

Estimate of RSD

$$= \sqrt{\{[(2.828)^2 + (0.000)^2 + (2.828)^2 + (5.657)^2}$$

$$+ (2.828)^2 + (5.657)^2 + (5.657)^2 + (2.828)^2]/8\}$$

$$= 4.000$$

Had all 16 batches been produced using the same formulation we could have expected a standard deviation of approximately 4.00 for the tensile strength measurements. The standard deviation of the measurements in Table 12.21 is, of course, much greater than 4.00 because of the variation due to the changes in the independent variables. We will now calculate effect estimates in an attempt to quantify the effects of the three independent variables.

The three-way table can, of course, be presented as three two-way tables (Table 12.23) to facilitate the calculation of effect estimates. Each entry in the two-way tables is the mean of two cell means from Table 12.22.

Table 12.23   Two-way tables of mean response values

(a)				(b)				(c)		
	z				w				w	
x	Low	High		x	Low	High		z	Low	High
Low	41	43		Low	42	42		Low	64	54
High	77	71		High	69	79		High	47	67

Using the row means, the column means and the diagonal means we can calculate the effect estimates as follows:

Main effect x $= \frac{1}{2}(77+71) - \frac{1}{2}(41+43) = 32$

Main effect z $= \frac{1}{2}(43+71) - \frac{1}{2}(41+77) = -2$

Main effect w $= \frac{1}{2}(54+67) - \frac{1}{2}(64+47) = 5$

Interaction xz $= \frac{1}{2}(41+71) - \frac{1}{2}(43+77) = -4$

Interaction xw $= \frac{1}{2}(42+79) - \frac{1}{2}(42+69) = 5$

Interaction zw $= \frac{1}{2}(64+67) - \frac{1}{2}(54+47) = 15$

Returning to the three-way table, Table 12.21, we can also calculate the three-variable interaction as:

Interaction xzw $= \frac{1}{2}\{[\frac{1}{2}(34+82) - \frac{1}{2}(60+52)]$

$- [\frac{1}{2}(50+76) - \frac{1}{2}(78+32)]\}$

$= -3$

The statistical significance of each effect estimate can be assessed by means of a *t*-test. We will first test the largest estimate, main effect x.

Null hypothesis – Tensile strength of the rubber is not dependent on the carbon black content.

Alternative hypothesis – Tensile strength of the rubber is dependent on the carbon black content.

$$\text{Test statistic} = \frac{\text{effect estimate}}{RSD/\sqrt{(p2^{n-2})}}$$

$$(\text{where } p = 2 \text{ and } n = 3)$$

$$= \frac{32.0}{4.00/\sqrt{4}}$$

$$= 16.0$$

Critical values – From the two-sided t-table with 8 degrees of freedom:

2.31 at the 5% significance level
3.36 at the 1% significance level
5.04 at the 0.1% significance level.

Decision – We reject the null hypothesis at the 0.1% level of significance.

Conclusion – We conclude that the tensile strength of the rubber is dependent on the carbon black content.

Each of the seven effect estimates can be subjected to the t-test. Continuing down the list we would conclude that:

(a) interaction zw is very highly significant (i.e. at 0.1%);
(b) main effect w is significant (i.e. at 5%);
(c) interaction xw is significant (i.e. at 5%);
(d) main effect z, interaction xz and interaction xzw are not significant.

In carrying out the above tests we are concerned with *statistical significance* rather than *practical importance*. We must now return to the world of reality and express our conclusions in the language of the research and development chemist. In the statement of the problem it was suggested that this experiment was a preliminary investigation which would provide information on which to base further experiments. Are we now in a position to specify treatment combinations which are worthy of investigation in our quest for increased tensile strength?

Starting with the main effects we see that x and w are both significant and that both of the effect estimates are *positive*. To increase tensile strength therefore we should adopt the *high* level of variable x and the *high* level of variable w. In practical terms we should use 40 parts per hundred of carbon black and we should use 75% natural rubber in the formulation. The significance of interaction xw directs our attention to Table 12.23(b), where we see that the high levels of both factors give the greatest value of tensile strength. This confirms what we have already established. The significance of interaction zw directs our attention to Table 12.23(c) which indicates that maximum tensile strength results from:

(a) *either* low level of variable z with low level of variable w.
(b) *or* high level of variable z with high level of variable w.

Since we have already decided that the high level of variable w is desirable, we will now adopt the high level of variable z, i.e. we will use the HLX accelerator.

In the light of the above conclusions, the research and development chemist might advance his investigation one stage further by carrying out a 2^2 factorial experiment using:

(a) 40 parts per hundred and 50 parts per hundred of carbon black;
(b) 75% and 85% of natural rubber;
(c) the HLX accelerator in all four trials.

This proposed experiment is, of course, just one of the many possibilities open to the chemist at this point. Whether he carries out this 2^2 factorial experiment or some alternative, he will be conscious of many *constraints* on his freedom to explore. Most of these constraints have not even been mentioned in this discussion.

12.10 More regression analysis

We will complete this chapter by using our regression package yet again. The data in Table 12.21 will be reanalysed so that we can compare the conclusions suggested by multiple regression analysis with the conclusions reached after using the t-test. Will we again decide that x, zw, xw and z are statistically significant?

To make the two analyses more comparable we will not specify a significance level for the regression package. The result will be that *all* of the independent variables are brought into the equation, starting with the *most* significant and finishing with the *least* significant. This is the strategy we have adopted in all of the regression analyses carried out in this chapter.

Table 12.24 Two replicates of a 2^3 experiment

X	Z	W	XZ	XW	ZW	XZW	y
−1	−1	−1	+1	+1	+1	−1	248
−1	−1	−1	+1	+1	+1	−1	252
−1	−1	+1	+1	−1	−1	+1	232
−1	−1	+1	+1	−1	−1	+1	232
−1	+1	−1	−1	+1	−1	+1	236
−1	+1	−1	−1	+1	−1	+1	232
−1	+1	+1	−1	−1	+1	−1	256
−1	+1	+1	−1	−1	+1	−1	248
+1	−1	−1	−1	−1	+1	+1	280
+1	−1	−1	−1	−1	+1	+1	276
+1	−1	+1	−1	+1	−1	−1	272
+1	−1	+1	−1	+1	−1	−1	280
+1	+1	−1	+1	−1	−1	−1	264
+1	+1	−1	+1	−1	−1	−1	256
+1	+1	+1	+1	+1	+1	+1	284
+1	+1	+1	+1	+1	+1	+1	280

The data fed into the computer contain values of the four measured variables:

(a) tensile strength (y);
(b) carbon black content (x);
(c) type of accelerator (z);
(d) % natural rubber (w).

The three independent variables (x, z, w) are first scaled and then four cross-product variables are generated to give the full set of data in Table 12.24.

The correlation matrix for the data in Table 12.24 is largely predictable and contains the highly desirable zero intercorrelations between the seven independent variables. As we would expect with such a matrix the sequence of regression equations shows no changes in the coefficients as variables are introduced. The equations are:

$y = 258.0 + 16.0X$	75.29% fit
$y = 258.0 + 16.0X + 7.5ZW$	91.84% fit
$y = 258.0 + 16.0X + 7.5ZW + 2.5XW$	93.68% fit
$y = 258.0 + 16.0X + 7.5ZW + 2.5XW + 2.5W$	95.51% fit
$y = 258.0 + 16.0X + 7.5ZW + 2.5XW + 2.5W - 2.0XZ$	96.69% fit
etc.	etc.

If we compare these equations with the earlier analysis we find substantial agreement. The coefficient of X (16.0) in the first regression equation is equal to exactly half of the effect estimate (32.0) calculated earlier. The former tells us that an increase in tensile strength of 16 units can be expected from a carbon black increase of 10 units. The latter tells us that a tensile strength increase of 32 units can be expected from the total change of 20 units of carbon black used in the experiment. Clearly it is not difficult to misinterpret the effect estimate from *either* of the two methods of analysis. Fortunately it is easier to interpret the regression equation after the independent variables $(X, W$ etc.) have been *descaled*. Before the descaling is carried out, however, we must decide which of the terms in the regression equation are statistically significant. This can be achieved by means of a sequence of t-tests based on the changes in percentage fit. You may recall that, in the previous chapter, we calculated the test statistic for this test using:

Test statistic =
$$\sqrt{\{[(\text{new \% fit} - \text{old \% fit}) \, (n - k - 1)]/(100 - \text{new \% fit})\}}$$

where k is the number of independent variables in the regression equation.

Using this formula again we get the test statistics in Table 12.25. From this table we see that the first independent variable (X) and the second independent variable (XW) are statistically significant at the 1% level of significance. The third and subsequent variables are *not* significant, even at the 5% level. Unfortunately this is *not* in agreement with the conclusion suggested by the t-tests which indicated that x, zw, xw and z were significant. What is the reason for this discrepancy?

Table 12.25 t-tests on the multiple regression equation

Independent variables in the equation	Test statistic	Critical values from t-table	
		5%	1%
X	6.53	2.14	2.98
X, ZW	5.13	2.16	2.01
X, ZW, XW	1.86	2.18	3.05
X, ZW, XW, W	2.18	2.20	3.10
X, ZW, XW, W, XZ	1.89	2.23	3.17

Consider the regression analysis after the third variable has entered the equation. The percentage fit is now 93.68%. Subtracting this figure from 100% we find that 6.32% of the variation in tensile strength remains unexplained. This 6.32% appears in the denominator of the test statistic where it is being used as a yardstick against which to compare the increase in percentage fit. It could be argued that the use of this yardstick does not give the third variable a fair chance of getting into the equation. Whilst part of the 6.32% is undoubtedly due to 'error' it is also possible that part is due to the effect of an independent variable which is not yet in the equation.

Our regression analysis, therefore, is more conservative than the analysis based on t-tests since the increase in percentage fit is being tested against an inflated residual. Greater comparability of the two approaches can be obtained if we use a regression package which is based on 'variable rejection' rather than 'variable selection'. Such a package would first fit an equation containing *all seven* of the independent variables $(X, Z, W, XZ, XW, ZW$ and $XZW)$ and the residual standard deviation from this equation would actually be equal to that calculated from Table 12.22 (i.e. 4.00). The least significant variable would then be tested against this residual standard deviation and rejected from the equation if the test statistic was less than the critical values. If this variable is rejected from the equation its sum of squares is included in the residual sum of squares before the next variable is tested. After a succession of rejections a final equation is reached and with some sets of data this equation would include a larger number of independent variables than the equation arrived at by a 'variable selection' method. In other cases the two methods would lead us to exactly the same conclusion.

12.11   Summary

In this chapter we have examined 2^2 and 2^3 factorial experiments. By means of such experiments we can estimate the main effects of the variables and the interactions between the variables. These estimates are independent of each other as the independent variables are not correlated with each other in a 2^n factorial experiment. We can therefore reach more positive conclusions than we were able to reach in the previous chapter when analysing the results of the plant manager's experiment.

We have seen that regression analysis can be used to analyse the data from a factorial experiment but its indiscriminate use can lead us astray especially if the independent variables have not been scaled. When a 2^n factorial experiment has been designed and carried out according to plan then it is probably better to avoid the use of regression analysis. If, on the other hand, the experiment is incomplete or does not match up to the plan for some reason, the use of a regression package might help us to salvage a great deal.

In the next chapter we will consider ways in which the size of a factorial experiment can be reduced without sacrificing the most important estimates.

Problems

(1) Dullness has been observed in recent batches of a particular dyestuff and a laboratory experiment has been called for in order to discover the cause of the dullness. The last ten batches that have been produced on the plant have brightness values in the range -4 to 0 with an average brightness of -2. It is important that the plant operating conditions be changed as soon as possible to obtain an average brightness of 0. (Brightness is measured on a scale from -7 to $+7$.)

It is decided that three variables will be included in the laboratory experiment the details of which are tabulated below:

Trial number	Speed of agitation	Reaction temperature (°C)	pH	Brightness
1	Fast	80	6	-2
2	Fast	90	8	0
3	Slow	90	6	-2
4	Fast	80	8	2
5	Slow	80	8	-4
6	Slow	80	6	0
7	Fast	90	6	-6
8	Slow	90	8	-4

(a) What is the name given to this type of experiment?
(b) What would you be able to estimate from the results of this experiment?
(c) Add 6 to the brightness values to eliminate the negative values.
(d) Put the results into a three-way table.
(e) Produce three two-way tables from the three-way table.
(f) Calculate estimates of the main effects and interaction effects from the two-way tables.
(g) Four earlier trials each with medium speed, 85 °C and a pH of 7 were found to have a standard deviation of brightness equal to 1.15. Use this as an estimate of the residual standard deviation to test the significance of the effect estimates.

(2) The data from Problem 1 were analysed using a multiple regression program. The actual values of temperature and pH were used and the speed was quantified as $0 = $ slow and $1 = $ fast. The interactions were generated by the program without coding the data. The following output was obtained, where s is speed, t is temperature, p is pH and y is brightness:

Correlation coefficients

	s	t	p	$s \times t$	$s \times p$	$t \times p$	$s \times t \times p$
y	0.20	−0.41	0.20	0.17	0.34	0.04	0.31

Correlation matrix

	s	t	p	$s \times t$	$s \times p$	$t \times p$	$s \times t \times p$
s	1.00	0.00	0.00	0.99	0.98	0.00	0.98
t		1.00	0.00	0.06	0.00	0.38	0.06
p			1.00	0.00	0.14	0.92	0.14
$s \times t$				1.00	0.98	0.02	0.98
$s \times p$					1.00	0.13	0.99
$t \times p$						1.00	0.15
$s \times t \times p$							1.00

Regression equations

$$y = 15.00 - 0.20t \qquad\qquad 16.67\%$$

$$y = 14.2 - 0.20t + 0.23\,(s \times p) \qquad\qquad 28.43\%$$

$$y = 6.2 - 0.10t - 0.19\,(s \times t) + 2.50\,(s \times p) \qquad\qquad 77.08\%$$

$$y = 11.5 - 0.05t - 1.40p - 0.30\,(s \times t) + 3.80\,(s \times p) \qquad\qquad 94.60\%$$

$$y = 71.0 - 0.75t - 9.90p - 0.30\,(s \times t) + 3.80\,(s \times p) \\ + 0.1\,(t \times p) \qquad\qquad 98.77\%$$

$$y = 76.0 - 10.0s - 0.8t - 10.0p - 0.2\,(s \times t) + 4.0\,(s \times p) \\ + 0.1\,(t \times p) \qquad\qquad 100.0\%$$

(a) Complete the following table which refers to values used by the computer in the multiple regression analysis:

Trial	s	t	p	$s \times t$	$s \times p$	$t \times p$	$s \times t \times p$
1	1	80	6	80	6	480	480
2		90	8				
3	0	90	6				
4		80	8				
5		80	8				
6		80	6				
7		90	6				
8		90	8				

(b) Examine the correlation matrix. Why are there so many high correlations?
(c) If the data were coded (-1 and $+1$), what correlation coefficient would be given by s and ($s \times t$)?
(d) Using the percentage fit statistic, carry out a significance test at each stage of the regression.
(e) If a 5% significance level had been stipulated as a criterion for cessation of the stepwise regression analysis, what conclusions would have been reached?
(f) Using the final multiple regression equation, calculate a predicted value for brightness with a fast agitator speed, a temperature of 90 and a pH of 8.
(g) Why was 100% fit obtained without the inclusion of the three-variable interaction?

─────────13─────────
Reducing the size of an experiment

13.1   Introduction

In the previous chapter we examined the 2^2 and 2^3 factorial experiments. These experiments are useful if we have two or three independent variables and we wish to explore only two values of each variable. You may recall that the plant manager, whose problems were discussed in Chapters 10 and 11, included *five* independent variables in his experiment and, furthermore, he used at least *three* values for each variable. As the plant manager only carried out ten trials (i.e. he produced ten batches) it is small wonder that his experiment lacked the admirable qualities that we find in a factorial experiment. On the other hand it would be out of the question to carry out an experiment involving 243 batches, which is the number required for a 3^5 factorial experiment.

In this chapter we will explore the possibility of getting *some* of the advantages of a factorial experiment without undertaking the *whole* of the experiment.

13.2   The design matrix

In the previous chapter we analysed the results from several factorial experiments and for each experiment we used *two* methods of analysis. The first was based on two-way tables and the second made use of a multiple regression package. Both methods have their advantages but there are other methods which are found to be even more useful when dealing with parts of factorial experiments as we shall see later in this chapter.

Prominent amongst these other methods of analysis is the very popular *Yates' technique*. Unfortunately the Yates' approach is expressed in a mathematical notation which has to be mastered before analysis can begin. We will not, therefore, discuss Yates' technique in this book, preferring instead to make use of what is known as the *design matrix*. This is basically very simple and involves very little notation; in fact we have already used several design matrices in the previous chapter without referring to them as such. One is reproduced as Table 13.1.

Table 13.1 below is very similar to Table 12.19. The main difference between

Table 13.1 2^3 experiment – design matrix and response vector

$$
\begin{array}{cccccccc}
X & Z & W & XZ & XW & ZW & XZW \\
\end{array}
$$

$$
\begin{bmatrix}
-1 & +1 & -1 & -1 & +1 & -1 & +1 \\
+1 & +1 & +1 & +1 & +1 & +1 & +1 \\
+1 & -1 & +1 & -1 & +1 & -1 & -1 \\
-1 & -1 & +1 & +1 & -1 & -1 & +1 \\
+1 & -1 & -1 & -1 & -1 & +1 & +1 \\
+1 & +1 & -1 & +1 & -1 & -1 & -1 \\
-1 & +1 & +1 & -1 & -1 & +1 & -1 \\
-1 & -1 & -1 & +1 & +1 & -1 & -1 \\
\end{bmatrix}
\qquad
\begin{bmatrix}
236 \\
284 \\
272 \\
232 \\
280 \\
264 \\
256 \\
248 \\
\end{bmatrix}
$$

the two is the separation of the *y* column from the other seven columns. By making this split we are distinguishing between the dependent variable (*y*) and the independent variables (*X*, *Z*, *W*, *XZ*, *XW*, *ZW* and *XZW*). To emphasize this split we refer to the *y* column as the *response vector* and to the other columns together as the *design matrix*. These names are very suitable as the design matrix is determined entirely by the *design* of the experiment and could be written down *before* the experiment was carried out, whereas the response vector contains the *results* of the experiment and is therefore not available until *after* the experiment is completed.

If we are contemplating carrying out a particular experiment we can set out the design matrix in advance and by examining its peculiarities we can determine whether or not the design meets our objectives. We will use the design matrix in this way later in the chapter but first we will make use of Table 13.1 to calculate the effect estimates which were obtained by two-way tables in the previous chapter.

To calculate the main effect of variable *x* we use the response vector and the *X* column of the design matrix as follows:

$$
\text{Main effect } x = (-236 + 284 + 272 - 232
$$

$$
+ 280 + 264 - 256 - 248)/4
$$

$$
= 32.0
$$

In this calculation the numbers are taken from the response vector and the signs are taken from the *X* column of the design matrix. The other six columns of the matrix can be used in a similar manner. For example, the *Z* column of the design matrix together with the response vector can be used to calculate an estimate of main effect *z*:

$$
\text{Main effect } z = (+236 + 284 - 272 - 232
$$

$$
- 280 + 264 + 256 - 248)/4
$$

$$
= 2.0
$$

Clearly we can also calculate estimates of the other main effect (*w*), the two-variable interactions (*xz*, *xw* and *zw*) and the three-variable interaction (*xzw*). In each case we would obtain exactly the same value as that calculated from the two-way tables in the previous chapter. The design matrix offers us a simple way of analysing the results of a factorial experiment but the main use to which we will put the design matrix is to compare alternative experiments *before* they have been carried out. We will pursue this in the next section.

13.3   Half replicates of a 2^n factorial design

In many situations it may not be possible to carry out a factorial experiment to investigate the factors in which we are interested. The prohibiting constraint may be the time or cost if we wish to examine several factors. It is unfortunately true that a 2^n factorial experiment requires a large number of trials if n is large, as we see in Table 13.2.

Table 13.2   The number of trials in 2^n factorial experiment

Number of factors (n)	2	3	4	5	6	7
Number of trials (2^n)	4	8	16	32	64	128

Another criticism of the factorial experiment is that it may give us estimates that we do not require. This is not true of the 2^2 and 2^3 experiments we have considered so far but would certainly be true of a 2^6 factorial experiment. Imagine that a researcher wished to investigate the effect of six factors on the impurity in a synthesized product. If he was prepared to limit each factor to only two levels the 2^6 factorial experiment would involve 64 trials which could be very time consuming and/or very expensive. To compensate the researcher for his labours what estimates would he be able to obtain from his experiment? They are rather numerous as the list below shows:

  6 main effects
15 two-factor interactions
20 three-factor interactions
15 four-factor interactions
  6 five-factor interactions
  1 six-factor interaction

It is *very* doubtful if all of these estimates would be required. Indeed it might be very difficult to find any meaning for a four-factor interaction if it was shown to be significant.

Since the 2^6 experiment is, on the one hand, too large and on the other hand, too productive of effect estimates it is natural to ask 'Can I carry out *part* of the 2^6 factorial experiment in order to get the estimates I require whilst losing those which I neither need nor understand?' The answer is 'Yes', and we can use the design matrix to help the researcher to select *which* part of the 2^6 factorial experiment he will carry out.

To illustrate the usefulness of the design matrix for this purpose let us return to the 2^3 factorial experiment first described in the previous chapter. Suppose that the research and development chemist carrying out the experiment had been forced to call a halt after *the first four trials* had been completed. Since the full series of eight trials constitutes a 2^3 factorial experiment we refer to these four trials as a *half replicate* of a 2^3 factorial experiment. (Similarly we would call any two of the eight trials, a quarter replicate.)

What can the research and development chemist salvage from his half replicate? Can he calculate estimates of the three main effects and the four interactions? The answers to these questions must lie in the top four rows of the design matrix in Table 13.1, which are reproduced in Tables 13.3 and 13.4 with the corresponding response values.

Table 13.3   Design matrix for a half replicate of the 2^3 experiment

X	Z	W	XZ	XW	ZW	XZW		y
−	+	−	−	+	−	+		236
+	+	+	+	+	+	+		284
+	−	+	−	+	−	−		272
−	−	+	+	−	−	+		232

Table 13.4   Three-way table for a half replicate of the 2^3 experiment

	Z			
	Low		High	
	W		W	
X	Low	High	Low	High
Low		232	236	
High		272		284

Tables 13.3 and 13.4 present the same information in alternative forms and either the design matrix or the three-way table could be used to calculate effect estimates. Using the *X* column of the design matrix we get:

$$\text{Main effect } x = (-236 + 284 + 272 - 232) \div 2$$
$$= 44$$

and using the *Z* column we can estimate the second main effect:

$$\text{Main effect } z = (+236 + 284 - 272 - 232) \div 2$$
$$= 8$$

When we come to estimate main effect *w* we notice that the *W* column contains three plusses and only one minus. Clearly this particular half replicate *is not balanced* as was the full 2^3 factorial experiment from which it was taken. Whereas each of the seven columns in Table 13.1 contains four pluses and four minuses there is no such balance of plus and minus signs in Table 13.3. This does not prevent us from calculating effect estimates but it does mean that we cannot use simplified formulae similar to those which are applicable to the full factorial experiment. With an unbalanced design we would have to work from first principles in calculating effect estimates and sums of squares. Furthermore we would get less precise estimates than could be obtained from a balanced design of the same size and these estimates would not be independent because of intercorrelation of the columns in the design matrix.

If this particular half replicate is lacking in the fine qualities that we found in the full factorial experiment, can we find a different half replicate which is more desirable? Thinking in terms of a balanced design, can we find four rows in Table 13.1 such that each of the seven columns will contain two plus and two minus signs? Unfortunately this is not possible. Indeed, it would be rather optimistic to expect *all* the benefits of a 2^3 design from only four trials. Perhaps we should concentrate on getting 'balanced' estimates of the three main effects. We can see in Table 13.4 the nature of the unbalance of this particular half replicate. Whilst the four response values are equally shared between the two rows they are *not* equally shared between the four columns. If we changed *one* of the four trials a more balanced design would result. The half replicate in Tables 13.5 and 13.6 is certainly better balanced than the one we have just considered.

Table 13.5 Design matrix for a second half replicate

X	Z	W	XZ	XW	ZW	XZW		y
−	+	−	−	+	−	+		236
+	+	+	+	+	+	+		284
+	−	+	−	+	−	−		272
−	−	−	+	+	+	−		248

Table 13.6 Three-way table for a second half replicate

	Z			
	Low		High	
	W		W	
X	Low	High	Low	High
Low	248		236	
High		272		284

From this half replicate we can calculate the following estimates, using either the design matrix or the three-way table:

$$
\begin{aligned}
\text{Main effect } x &= 36 \\
\text{Main effect } z &= 0 \\
\text{Main effect } w &= 36 \\
\text{Interaction } xz &= 12 \\
\text{Interaction } xw &= ?? \\
\text{Interaction } zw &= 12 \\
\text{Interaction } xzw &= 0
\end{aligned}
$$

You will notice in the above list that there is no estimate for interaction xw. It is not possible to calculate an estimate since the XW column of the design matrix contains four plus signs. You will also notice that the six estimates we *have* calculated fall into three pairs, i.e. we have the same value for main effect x and for main effect w, etc. This is not just a coincidence, as Table 13.5 shows. In the design matrix the X column and the W column are *identical*. Similarly, the Z column and the XZW column are identical. We summarize this situation by saying that:

xw is the *defining contrast*

$\left.\begin{array}{l} (x \text{ and } w) \\ (z \text{ and } xzw) \\ (xz \text{ and } zw) \end{array}\right\}$ are *alias pairs*

The two effects which constitute any alias pair are inseparable from each other. The calculated estimate of 36, for the (X and W) alias pair tells us that a change in tensile strength of 36 kg/cm^2 can be attributed to the change made to variable x (carbon black content) *or* to the change made to variable w (% natural rubber) *or* to both. As you can see in Table 13.5, the X column and the W column are identical. An experiment in which a change in one variable is *always* accompanied by a change in a second variable will never allow us to separate the effect of the two variables.

Since the least important of all the seven estimates is interaction XZW we will choose a half replicate which has interaction XZW as the defining contrast. To do this we refer to Table 13.1 and select the four trials which have a plus sign in the XZW column. This third example of a half replicate is set out in Tables 13.7 and 13.8.

Table 13.7   Design matrix for a third half replicate

X	Z	W	XZ	XW	ZW	XZW	y
−	+	−	−	+	−	+	236
+	+	+	+	+	+	+	284
−	−	+	+	−	−	+	232
+	−	−	−	−	+	+	280

Table 13.8   Three-way table for a third half replicate

	Z			
	Low		High	
	W		W	
X	Low	High	Low	High
Low		232	236	
High	280			284

Without calculating effect estimates we can see which effects will be aliased with each other by spotting pairs of identical columns in the design matrix. In Table 13.7 we see that:

Interaction *xzw* is the defining contrast
Main effect *x* is aliased with interaction *zw*
Main effect *z* is aliased with interaction *xw*
Main effect *w* is aliased with interaction *xz*.

With this half replicate, then, we could estimate main effect *x* *if* we were able to assume that interaction *zw* did not exist, and we could estimate main effect *z* if we were able to assume that there was no interaction between variable *x* and variable *w*. These assumptions would not be made on statistical grounds, of course. One wonders if there are *any* situations in which such assumptions could reasonably be made and for this reason the half replicate of a 2^3 experiment is not a widely used design. Though we have used a 2^3 factorial experiment to illustrate the ideas underlying *fractional replication*, the practical benefits of the technique will be enjoyed by the researcher who wishes to investigate the effect of *four or more factors*.

Table 13.9 Design matrices for 2^2, 2^3 and 2^4 factorial experiments

X	Z	XZ	W	XW	ZW	XZW	V	XV	ZV	XZV	WV	XWV	ZWV	XZWV
−	−	+	−	+	+	−	−	+	+	−	+	−	−	+
+	−	−	−	−	+	+	−	−	+	+	+	+	−	−
−	+	−	−	+	−	+	−	+	−	+	+	−	+	−
+	+	+	−	−	−	−	−	−	−	−	+	+	+	+
−	−	+	+	−	−	+	−	+	+	−	−	+	+	−
+	−	−	+	+	−	−	−	−	+	+	−	−	+	+
−	+	−	+	−	+	−	−	+	−	+	−	+	−	+
+	+	+	+	+	+	+	−	−	−	−	−	−	−	−
−	−	+	−	+	+	−	+	−	−	+	−	+	+	−
+	−	−	−	−	+	+	+	+	−	−	−	−	+	+
−	+	−	−	+	−	+	+	−	+	−	−	+	−	+
+	+	+	−	−	−	−	+	+	+	+	+	−	−	−
−	−	+	+	−	−	+	+	−	−	+	+	−	−	+
+	−	−	+	+	−	−	+	+	−	−	+	+	−	−
−	+	−	+	−	+	−	+	−	+	−	+	−	+	−
+	+	+	+	+	+	+	+	+	+	+	+	+	+	+

The whole of Table 13.9 constitutes a design matrix for a 2^4 factorial experiment. The sections which are boxed off constitute design matrices for a 2^3 experiment and a 2^2 experiment. To obtain the design matrix for a half replicate of a 2^4 factorial experiment we select eight rows of Table 13.9, using the full width of the table. If we want a half replicate in which interaction *xzwv* is the defining contrast we select those eight rows which have a plus sign in the last column. Careful examination of these eight rows would indicate that:

Main effect x   was aliased with interaction *zwv*
Main effect z   was aliased with interaction *xwv*
Main effect w was aliased with interaction *xzv*
Main effect v   was aliased with interaction *xzw*
Interaction *xz* was aliased with interaction *wv*
Interaction *xw* was aliased with interaction *zv*
Interaction *xv* was aliased with interaction *zw*.

With this particular half replicate of a 2^4 factorial design we can estimate each main effect if we can assume that the aliased three-factor interaction is negligible. We can also estimate a two-factor interaction if we are able to assume that the other two-factor interaction, with which it is aliased, does not exist.

When we have five or more factors a half replicate of the 2^n factorial experiment can yield estimates of all the main effects and two-factor interactions which are not aliased with each other. In a half replicate of a 2^5 design, with the five-variable interaction chosen as the defining contrast, we find that:

(a) each of the five main effects is aliased with a four-factor interaction;
(b) each of the ten two-factor interactions is aliased with a three-factor interaction.

13.4   Quarter replicates of a 2^n factorial design

If we wish to investigate *many* independent variables, even a half replicate of the full factorial experiment may be too large. In this case we could consider carrying out only a *quarter* of the 2^n trials. Clearly a quarter replicate will yield much less information than would a half replicate but we may, nonetheless, be able to estimate the main effects and the two variable interactions.

To illustrate certain points let us consider the first two trials of the 2^3 factorial experiment set out in Table 13.1. These two trials constitute a quarter replicate of the full 2^3 experiment and are summarized in Table 13.10.

Table 13.10   Design matrix for a quarter replicate of the 2^3 experiment

X	Z	W	XZ	XW	ZW	XZW		y
−1	+1	−1	−1	+1	−1	+1		236
+1	+1	+1	+1	+1	+1	+1		284

You will notice that in the design matrix of Table 13.10 three of the seven columns contain two plus signs. It is not possible therefore to estimate the three effects which correspond to these columns and we find that the quarter replicate of the 2^3 experiment has *three defining contrasts, z, xw* and *xzw*. The other four columns of the design matrix are identical and we can therefore calculate only one effect estimate:

$$-236 + 284 = 48.0$$

Thus the four effects (x, w, xz and zw) form an alias group and we are unable to say which one of the four has given rise to the increase in tensile strength of 48 units.

Clearly a quarter replicate of a 2^3 factorial experiment is quite useless but this particular example has served to illustrate two important points that are true of *all* quarter replicates:

(a) There are three defining contrasts.
(b) Each effect will fall into an alias group containing three other effects.

A quarter replicate of a 2^4 factorial experiment would consist of four trials compared with the 16 trials of a full 2^4 experiment. It would have three defining contrasts and three alias groups each containing four effects. No matter which four trials (of the possible 16) were carried out it would not be possible to separate even the four main effects.

A quarter replicate of a 2^5 experiment would consist of eight trials compared with the 32 trials in a full 2^5 experiment. The quarter replicate would have three defining contrasts and there would be seven groups of alias pairs, each containing four effects. It could be arranged that each of the five main effects was in a separate alias group but it would not be possible to ensure that each of the ten two-variable interactions was separated in this way.

If we wished to investigate the effect of *seven* independent variables then a quarter replicate of the full 2^7 experiment would require 32 trials. From the results of such an experiment we would be able to estimate the seven main effects and the 21 two-variable interactions independently. The $\frac{1}{4} \times 2^7$ design might therefore be more attractive than a full 2^7 design, to a researcher who wished to estimate only the main effects and the two-variable interactions.

13.5 A useful method for selecting a fraction of a 2^n factorial experiment

We have seen that *inspection of the design matrix* is a very powerful tool for evaluating the effectiveness of any experimental design before the experiment is carried out. Intercorrelation of the columns of the matrix indicate deficiencies which it may be possible to eradicate before it is too late. When the experiment is a fraction of a 2^n factorial we may get correlations of $+1.0$ or -1.0 and the effects then fall into alias groups. Inspecting the design matrix enables us to identify the alias groups and the defining contrasts.

The design matrix is not so useful to someone trying to design an experiment as to someone evaluating a design already written down. This point would become very clear if you pored over the 127 columns of a 2^7 design matrix trying to extract a suitable quarter replicate. At the design stage it is easier to use the following method which is carried out in two steps:

(a) Choose defining contrast(s).
(b) Generate alias groups by 'multiplying' each effect by each defining contrast.

Suppose for example that we wish to design a half replicate of a 2^3 factorial experiment in which the three independent variables are represented by X, W and Z. A half replicate has *only one* defining contrast so we will choose the three-factor interaction XWZ. Now we must 'multiply' each of the main effects (X, Z, W) and each of the interaction effects (XZ, XW, ZW, XZW) by the defining contrast. In performing these multiplications we use the normal rules of algebra plus the additional rule that $X^2 = 1$, $W^2 = 1$ and $Z^2 = 1$. The multiplication proceeds as follows:

$$
\begin{aligned}
X \times XZW &= X^2ZW &&= ZW &&(\text{as } X^2 = 1) \\
Z \times XZW &= XZ^2W &&= XW &&(\text{as } Z^2 = 1) \\
W \times XZW &= XZW^2 &&= XZ &&(\text{as } W^2 = 1) \\
XZ \times XZW &= X^2Z^2W &&= W \\
XW \times XZW &= X^2ZW^2 &&= Z \\
ZW \times XZW &= XZ^2W^2 &&= X \\
XZW \times XZW &= X^2Z^2W^2 &&= ?
\end{aligned}
$$

The result of the first multiplication ($X = ZW$) tells us that main effect X and interaction ZW constitute an alias pair. This same indication is given by the sixth multiplication so we have clearly done more work than was necessary. The other two alias pairs are seen to be Z with XW, and W with XZ.

Having listed the alias pairs it is now easy to write out the design matrix for the four trials which constitute the half replicate. We know that each column except XZW must contain two plus signs and two minus signs. We also know that the X, W and Z columns must have zero correlation with each other. If we write $+ + - -$ for X, $+ - + -$ for Z and $- + + -$ for W the other columns can be obtained by multiplication.

As a second example let us design a quarter replicate of a 2^4 factorial experiment using the letters A, B, C and D to represent the independent variables. For this exercise we will, of course, require *three* defining contrasts. We choose *two* of the higher interactions as our first two defining contrasts and then we obtain the third by multiplication. If we choose the three-factor interaction ABC as the first defining contrast and the three-factor interaction ACD as the second then by multiplication we obtain the third:

$$ABC \times ACD = A^2BC^2D = BD$$

We must now multiply each main effect and interaction by all three defining contrasts. Starting with main effect A we get:

$$A \times ABC = A^2BC = BC$$
$$A \times ACD = A^2CD = CD$$
$$A \times BD = ABD$$

Thus A, BC, CD and ABD form an alias group. Multiplying main effect B and then main effect C in the same way gives us the other two alias groups:

$$B, AC, D, ABCD \qquad \text{and} \qquad C, AB, BCD, AD$$

13.6   A hill-climbing approach to optimization

There are situations in which estimates of main effects and interactions are *not* of prime interest. It is possible that a researcher might wish to carry out a *series* of experiments. From the *first* experiment he simply requires an indication of what values of the independent variables he should use in his *second* experiment. The researcher is adopting what is known as a *hill-climbing* strategy, by means of which he hopes to eventually reach the summit where he will find the maximum possible yield. (Alternatively he might wish to travel down hill to the minimum impurity.)

The hill-climbing approach, or response-surface methodology as it is sometimes called, is dealt with fully in Davies (1978) and Box, Hunter and Hunter (1978). With this approach one would use only *two* values for each independent variable in the earlier experiments and fractions of 2^n factorial designs are often employed. In later experiments one would use *three* values for each independent variable and a 3^n factorial design might be useful. Clearly 3^n may be very much bigger than 2^n as Table 3.11 shows, but it is possible that several independent variables will have been eliminated before the final experiment is reached.

Table 13.11 2^n and 3^n factorial designs

Number of independent variables	2	3	4	5	6	7
Number of trials in a 2^n design	4	8	16	32	64	128
Number of trials in a 3^n design	9	27	81	243	729	2187

Designs of 3^n factorial are very important but shortage of space prevents their discussion within this book. Equally important is the general factorial design in which the variables have different numbers of levels. For example a $5 \times 4 \times 2$ factorial experiment would involve three independent variables with the first having five levels, the second having four levels and the third having only two levels. The whole experiment would require 40 trials (i.e. $5 \times 4 \times 2$). The analysis of the results of this experiment could be based on a three-way table which would be transformed into three two-way tables. Alternatively, multiple regression analysis could be used.

For a thorough coverage of 3^n factorial designs and general factorial designs the reader is referred to Davies (1978) or Box, Hunter and Hunter (1978). We will now discuss a complication which often arises in the design of experiments and which will be of interest when we return to the plant manager's experiment in the next chapter.

13.7   Blocking and confounding

When we come to consider the implementation of a planned experiment we may realize that the whole experiment cannot be carried out under homogeneous conditions. Perhaps we have designed a 2^4 factorial experiment but we find that a consignment of raw material is only sufficient for eight trials. Alternatively we might wish to carry out a 2^3 experiment but heat treatment of the eight test pieces cannot be carried out in one run because the oven will only accommodate six pieces. Because of such constraints a planned experiment may have to be carried out in two or more blocks. The subdivision of the whole experiment into suitable blocks must be done with care if the results are not to be ruined by block to block variation.

To illustrate the problems involved in blocking we will return to the 2^3 factorial experiment discussed in Chapter 12. The results of this experiment were put into a three-way table which is reproduced as Table 13.12.

Table 13.12   Results of a 2^3 experiment (from Table 12.17)

			z	
		Low		*High*
		w		*w*
x		*Low*	*High*	*Low*	*High*
Low		48	32	36	56
High		80	72	64	84

You may recall that the results in the three-way table were averaged in pairs to obtain the entries in three two-way tables. These are also reproduced here, as Table 13.13.

Table 13.13   (from Table 12.18)

x	*z* Low	*z* High	*x*	*w* Low	*w* High	*z*	*w* Low	*w* High
Low	40	46	Low	42	44	Low	64	52
High	76	74	High	72	78	High	50	70

To complete the analysis we calculated estimates of the three main effects and the four interactions as follows:

$$
\begin{aligned}
\text{Main effect } x &= \tfrac{1}{2}(74+76) - \tfrac{1}{2}(46+40) = 32 \\
\text{Main effect } z &= \tfrac{1}{2}(74+46) - \tfrac{1}{2}(76+40) = 2 \\
\text{Main effect } w &= \tfrac{1}{2}(78+44) - \tfrac{1}{2}(72+42) = 4 \\
\text{Interaction } xz &= \tfrac{1}{2}(74+40) - \tfrac{1}{2}(76+46) = -4 \\
\text{Interaction } xw &= \tfrac{1}{2}(78+42) - \tfrac{1}{2}(72+44) = 2 \\
\text{Interaction } zw &= \tfrac{1}{2}(70+64) - \tfrac{1}{2}(50+52) = 16 \\
\text{Interaction } xzw &= \tfrac{1}{2}\{(\tfrac{1}{2}(36+84) - \tfrac{1}{2}(64+56)) \\
&\quad -(\tfrac{1}{2}(48+72) - \tfrac{1}{2}(80+32))\} \\
&= -2
\end{aligned}
$$

Suppose that the raw material used in the manufacture of the eight samples of rubber is purchased in 50 kg bags and that 10 kg are required in the manufacture of each sample. Obviously we will need to use two bags of raw material, but our past experience suggests that there may be considerable variation from bag to bag. If we were to manufacture four samples from one bag and the other four samples from a second bag, what effect would this have on our conclusions? The answer to this question depends upon which of the eight trials are allocated to bag 1 and which to bag 2. Suppose that the experiment is implemented as shown in Table 13.14.

Table 13.14 2^3 factorial experiment using two bags of raw material

		z		
	Low		High	
	w		w	
x	Low	High	Low	High
Low	Bag 1	Bag 2	Bag 2	Bag 2
High	Bag 2	Bag 1	Bag 1	Bag 1

The reader may see immediately that the allocation of trials to bags in Table 13.14 has been done very badly. Three of the trials in the top row (i.e. at the low level of x) have been allocated to bag 2 whereas only one of the trials in the bottom row was allocated to this bag. If the response, the tensile strength of the rubber, is affected by the difference between the two bags then we would surely expect our estimate of main effect x to be in error. This can be easily demonstrated if we assume that the effect of the difference between the two bags of raw material is to increase by an unknown quantity, X, the tensile strength of the four samples made from bag 1. The results of the experiment would then be as in Table 13.15.

The entries in Table 13.15 are exactly the same as those in Table 13.12 except

Table 13.15   The effect of using two bags of raw material

		z		
		Low		High
		w		w
x	Low	High	Low	High
Low	48 + X	32	36	56
High	80	72 + X	64 + X	84 + X

for the addition of X, the unknown change in tensile strength. When we average the results in Table 13.15 to produce three two-way tables we find that X appears there also (Table 13.16).

Table 13.16   The effect of using two bags of raw material

x	z Low	High	x	w Low	High	z	w Low	High
Low	$40 + \dfrac{X}{2}$	46	Low	$42 + \dfrac{X}{2}$	44	Low	$64 + \dfrac{X}{2}$	$52 + \dfrac{X}{2}$
High	$76 + \dfrac{X}{2}$	$74 + X$	High	$72 + \dfrac{X}{2}$	$78 + X$	High	$50 + \dfrac{X}{2}$	$70 + \dfrac{X}{2}$

We can see by the lack of balance in the two-way tables that several of the effect estimates will be affected by the difference between bags. Completing the calculations we find that four of the estimates are affected whilst the other three are in agreement with the estimates calculated in the previous chapter.

$$\text{Main effect } x = \tfrac{1}{2}\left(74 + X + 76 + \frac{X}{2}\right) - \tfrac{1}{2}\left(46 + 40 + \frac{X}{2}\right) = 32 + \frac{X}{2}$$

$$\text{Main effect } z = \tfrac{1}{2}(46 + 74 + X) \quad - \tfrac{1}{2}\left(40 + \frac{X}{2} + 76 + \frac{X}{2}\right) = 2$$

$$\text{Main effect } w = \tfrac{1}{2}(78 + X + 44) \quad - \tfrac{1}{2}\left(72 + \frac{X}{2} + 42 + \frac{X}{2}\right) = 4$$

$$\text{Interaction } xz = \tfrac{1}{2}\left(74 + X + 40 + \frac{X}{2}\right) - \tfrac{1}{2}\left(76 + \frac{X}{2} + 46\right) = -4 + \frac{X}{2}$$

$$\text{Interaction } xw = \tfrac{1}{2}\left(42 + \frac{X}{2} + 78 + X\right) - \tfrac{1}{2}\left(72 + \frac{X}{2} + 44\right) = 2 + \frac{X}{2}$$

$$\text{Interaction } zw = \tfrac{1}{2}\left(64 + \frac{X}{2} + 70 + \frac{X}{2}\right) - \tfrac{1}{2}\left(50 + \frac{X}{2} + 52 + \frac{X}{2}\right) = 16$$

$$\text{Interaction } xzw = \tfrac{1}{2}\{[\tfrac{1}{2}(36 + 84 + X) - \tfrac{1}{2}(64 + X + 56)]$$
$$- [\tfrac{1}{2}(48 + X + 72 + X) - \tfrac{1}{2}(80 + 32)]\} = -2 - \frac{X}{2}$$

Obviously the arrangement suggested by Table 13.14 constitutes a bad design, as we anticipated. Is it possible to allocate the eight trials to the two bags in such a way that our estimates of all three main effects will be unaffected by the difference between bags? It certainly is, provided that we follow the line of thought which helped us to obtain useful half replicates in the first half of this chapter. The allocation specified in Table 13.17 is much better balanced than the one we have just considered and the reader will not be surprised to learn that this new arrangement produces much better estimates.

Table 13.17 A better allocation of trials to bags

	z			
	Low		*High*	
	w		*w*	
x	*Low*	*High*	*Low*	*High*
Low	Bag 2	Bag 1	Bag 1	Bag 2
High	Bag 1	Bag 2	Bag 2	Bag 1

If we again assume that the effect of using two bags of raw material is simply to increase the tensile strengths of the four bag 1 samples by X then we can calculate the effect estimates. By assuming that the *level* of tensile strength is changed but the effect of the independent variables is not changed we are in effect postulating that there is no interaction between the three independent variables (x, z, w) and the raw material. Working through the calculations we would find that only *one* estimate was affected by having carried out the experiment in two blocks. Furthermore it is the least important estimate, interaction xzw, that is in error, whilst the three main effects and the three two-variable interactions are completely unaffected by the blocking. We summarize this happy situation by saying that the *difference between blocks is confounded with interaction xzw*.

In the first blocked design that we considered (Table 13.14) the block effect was *partially confounded* with several effects. When splitting a 2^n factorial design into two blocks we can always arrange that the block effect will be confounded with one specified effect and we would usually choose the highest order interaction. If it were necessary to split a 2^n factorial experiment into four blocks then there would be three block effects to confound. Clearly there are strong similarities between confounded effects and the defining contrasts that arose in our discussion of fractional replicates.

13.8 Summary

In this chapter we have discussed *fractions* of 2^n factorial experiments. These can be particularly useful when we wish to investigate several independent variables and we are prepared to use only two values for each variable but we are not willing to carry out the full 2^n factorial. Clearly a half replicate (or a quarter replicate) cannot offer all the benefits of a full factorial experiment, but it may give us what we need most. This is particularly true if our immediate need is an indication of the values we should use in our *next experiment*.

Many researchers believe that in some situations it is more efficient to proceed by means of a series of small experiments rather than by hoping to solve all of one's problems in one grand design. If this sequential approach is adopted the earlier experiments could well be fractions of 2^n factorials. When the optimum region has been located it may be necessary to use three levels for the independent variables in the final design and the use of 3^n factorials is quite common.

We have also discussed, rather briefly, the effect of splitting a 2^n factorial experiment into two or more blocks. This course of action is often forced upon us by the constraints within which the experiment must be carried out. We will encounter blocking problems again in the next chapter and will examine how blocks can be taken into account when using multiple regression analysis.

Having spent two chapters studying some of the principles underlying experimental designs we can now return to the problem which we set aside at the end of Chapter 11. We will devote the next chapter to advising the plant manager on where he went wrong, what he should have done, and what he might now do if he wishes to reduce the impurity in future batches of pigment.

Problems

(1) It is believed that the viscosity of batches of polymer are affected by four variables: pressure (A), temperature (B), catalyst concentration (C) and reaction time (D). An experiment is suggested with the following design matrix, each variable having two levels represented by '+' and '−'.

Trial number	A	B	C	D
1	−	−	−	+
2	+	−	+	−
3	−	+	−	−
4	+	+	+	+

(a) What fraction of a full factorial design is this experiment?
(b) By multiplying the appropriate columns, generate the signs for all two-, three- and four-variable interactions to give the full design matrix.
(c) Examine the full design matrix and decide which are the three defining contrasts.
(d) Form alias groups of main effects and interactions.

(2) The chemist who carried out the 2^3 factorial experiment in Problem 1 of Chapter 12 wishes to design a further experiment. The purpose of this new investigation is twofold:

(i) To confirm the significance of the speed × pH interaction which was suggested by the first experiment.
(ii) To assess the effect of changing the concentration of the stabilizer.

Stabilizer concentration was not included in the first experiment, but the chemist suspects that it could be significant as a main effect and he has good reason to believe that stabilizer concentration will interact with temperature. It is also possible that stabilizer concentration will interact with pH and with speed of agitation.

The chemist decides that he will represent the four variables by the symbols A, B, C and D as follows:

A is agitation speed
B is pH
C is temperature
D is stabilizer concentration.

From the results of the new experiment it is essential that he should be able to estimate:

Main effects A, B and D;
Two-variable interactions AB, AD, BD and CD.

He is prepared to assume that main effect C and interactions AC and BC are negligible provided he uses the same levels for variables A, B and C that were used in the first experiment.

(a) Can the needs of the chemist be satisfied by a half replicate of a 2^4 factorial experiment?
(b) If the answer to (a) is 'yes', list the eight treatment combinations that you would advise him to use.

(3) It was suggested in Chapter 11 that your ability to make use of multiple regression analysis would be increased if you were able to understand the terminology used by computer packages and by statisticians. You will now realize that it is wise to consult a statistician or a good text *before* an investigation is carried out rather than *after* the data have been collected. Obviously, any such consultation will be enriched if the scientist has an appreciation of the terminology used in Chapters 12 and 13. Attempting to insert the missing words in the following passage will enable you to evaluate your ability to communicate in this medium.

In Chapter 12 we discussed 2^n factorial experiments in which we have (1)   independent variables (or factors), each having

(2)   values or levels. First we compared the 2^2 factorial experiment, which consists of four trials, with a classical, 'one factor at a time' experiment based on three trials. The advantages of the factorial experiment were:

(a) It gave more precise estimates because each effect estimate was based on the results of (3)   trials whereas each estimate in the classical experiment was based on only (4)   results;

(b) From the results of the 2^2 factorial it was possible to estimate not only the two main effects but also the (5)   effect.

The statistical significance of each effect estimate was checked by means of a (6)   . In order to calculate the test statistic in such a test we need to know the effect estimate, of course, and we also need an estimate of the (7)   . Unfortunately a 2^n factorial experiment does not yield such an estimate. By carrying out two replicates of a 2^2 factorial experiment, which requires (8)   trials in all, we can calculate an estimate of the residual standard deviation with (9)   degrees of freedom, whilst three replicates would give an estimate of the residual standard deviation with (10)   degrees of freedom.

The results of a 2^2 factorial experiment can be analysed by fitting multiple regression equations. If we call the response y and the two independent variables x and z, we can account for the two main effects by fitting the equation (11)   but to account for the interaction well we must fit (12)   . The extra variable in the equation is generated by multiplying each x value by the corresponding z value and is known as a (13)   variable. Unfortunately this generated variable will probably be highly (14)   with either or both of the measured variables, x and z, unless we (15)   these measured variables before multiplying. Provided this precaution has been taken we will have a perfect correlation matrix in which the correlation between any two independent variables is equal to (16)

When we have a large number of independent variables, even if we restrict each to two values, the number of trials required in a factorial experiment may be greater than we are prepared to carry out. Alternatively it may not be possible to carry out such a large number of trials under homogeneous conditions. We could decide, of course, to carry out only four of the eight trials which constitute a 2^3 factorial experiment. This would be known as a half (17)   and the results of the four trials could be used to calculate three effect estimates. If we had chosen the four trials such that the X column of the design matrix contained four $+$ signs then main effect X would be known as the (18)   The other two main effects and the four interaction effects would fall into three (19)   . Main effect Z would be aliased with interaction XZ, main effect W would be aliased with (20) whilst interaction XZW would be aliased with (21)   . Obviously this would be a bad half replicate for it would be impossible to estimate

(22)   and the estimate of main effect Z would only be useful if we were very confident that (23)   did not exist.

Similar problems arise if we decide to carry out a full 2^n factorial experiment but, because of constraints, we have to split it into two blocks. If the effect of blocking is simply to increase (or decrease) the response in one block by a fixed amount but not to change the effects of the independent variables we would say that there was no (24)   between blocks and independent variables. In the absence of such an interaction we could estimate all the effects except one and we would say that this particular effect was (25)   with the difference between blocks.

─────14─────
Improving a bad experiment

14.1 Introduction

In the last two chapters we have examined several experimental designs and highlighted some of the principles which must be respected if valid conclusions are to be drawn from experimental data. We have also discussed the possibility of carrying out only a fraction of a factorial experiment in order to reduce the number of trials whilst still being able to estimate the important effects.

Perhaps we are now in a position to return to the problems of the plant manager which we set aside at the end of Chapter 11. We have established that the experiment he carried out was less than perfect. Two of his independent variables, weight of special ingredient (x) and feedrate (z), were so very highly correlated ($+0.97$) that the main effects, x and z, almost constituted an alias pair and it was quite impossible to decide whether the observed reduction in impurity was caused by one or the other.

The plant manager was, of course, operating under severe constraints. There were strict limits to *how far* he could deviate from normal operating conditions and to *how many* batches he could include in any experimental series. Clearly these constraints cannot be ignored when we offer advice to the plant manager.

In this chapter we will specify operating conditions that could be used in a series of eight batches which will extend the plant manager's original experiment. The results from all 18 batches will then be analysed by means of multiple regression analysis and a strategy will be formulated for future production of this pigment.

Before we attempt this salvage operation, however, we will consider what the plant manager *should* have done in the first place as an alternative to the ineffective experiment that he actually did carry out.

14.2 An alternative first experiment

A useful starting point in any activity is a clear statement of objectives. When the activity under consideration involves the production of ten batches with operating conditions changing from batch to batch then the neglect of objectives is extremely unwise. On the other hand a clear statement of grandiose objectives will not in itself produce results. If, therefore, the plant manager hopes to investigate five independent variables in such a way that estimates of all main effects, quadratic effects and interactions can be obtained, within a strict limit of ten trials, then he is doomed to disappointment.

Let us start by considering the smallest possible experiment. With five independent variables, six trials are needed in order to obtain unambiguous estimates of all main effects. Such a small experiment would not give an estimate of residual variance and would not, therefore, enable the statistical significance of the estimates to be tested. With ten trials, however, we should be able to estimate five main effects *and* obtain an estimate of residual variance with four degrees of freedom.

> The plant manager's main objective, then, should be to estimate the five main effects.

It is possible that one or more of the five main effects will be shown to be not significant and this would release extra degrees of freedom for the estimate of residual variance. If the plant manager uses regression analysis to analyse the results and he ends up with only two significant main effects plus a residual variance with seven degrees of freedom then he might wish to introduce a quadratic term or a cross-product term into the equation. As a secondary objective, then, it is desirable to be able to estimate interaction effects and/or curvature of the main relationships. By pursuing this secondary objective we must be very careful not to jeopardize the primary objective.

The constraints within which we must operate are concerned with the number of trials and the range of variation of the independent variables. When choosing values for those variables we will stay within the limits used in the original experiment. These are given in Table 14.1

Table 14.1   The original experiment

Independent variable	Values used
Weight of special ingredient (x)	0, 1, 1, 2, 3, 3, 4, 5, 5, 6
Catalyst age (w)	1, 2, 3, 4, 5, 6, 7, 8, 9, 10
Feedrate (z)	1, 2, 2, 3, 3, 5, 5, 6, 6, 7
Temperature (t)	1, 1, 1, 2, 2, 2, 2, 3, 3, 3
Agitation speed (s)	2, 2, 2, 2, 3, 3, 3, 4, 4, 4

The proliferation of values displayed in Table 14.1 is a luxury we cannot afford if our experiment is limited to only ten batches. To estimate linear main effects and interactions we need only *two* values for each variable. If we wish to consider quadratic effects the number of values will need to be increased to *three*, but any further increase is neither necessary nor desirable. Catalyst age (w) is an exception as this variable will increase in value in steps of one from batch to batch.

With these thoughts in mind let us make a start. Perhaps the first step should be to choose a 'standard' design of a suitable size and then modify this to suit our purpose. Within the upper limit of 10 trials we have:

(a) a 1/27 replicate of a 3^5 factorial experiment, requiring nine trials;
(b) a 1/4 replicate of a 2^5 experiment, requiring eight trials;
(c) a 1/9 replicate of a 3^4 experiment, requiring nine trials;
(d) a 1/2 replicate of a 2^4 experiment, requiring eight trials.

To make use of the last two experiments in the list we would set aside catalyst age (w) and consider only the other four independent variables. The fifth variable (w) would re-enter the experiment when the trials were carried out in random order. Using this approach we could, purely by chance, find a high correlation between catalyst age and one of the other four variables. This could, of course, be checked before the experiment was carried out.

Whilst all four alternatives listed are worthy of consideration we will concentrate on the 1/4 replicate of a 2^5 factorial experiment. To obtain a suitable quarter we could write out the design matrix for a full 2^5 experiment and then, using three defining contrasts, select 8 of the 32 rows. As an alternative we will set out the design matrix of a 2^3 factorial experiment then we will add two more columns generated by multiplication. The result is shown in Table 14.2.

Table 14.2 A 1/4 replicate of a 2^5 factorial experiment

Trial	Variable				
	x	w	z	$t(=xw)$	$s(=xz)$
1	-1	-1	-1	$+1$	$+1$
2	$+1$	-1	-1	-1	-1
3	-1	$+1$	-1	-1	$+1$
4	$+1$	$+1$	-1	$+1$	-1
5	-1	-1	$+1$	$+1$	-1
6	$+1$	-1	$+1$	-1	$+1$
7	-1	$+1$	$+1$	-1	-1
8	$+1$	$+1$	$+1$	$+1$	$+1$

In Table 14.2 the entries in the t column have been obtained by multiplying corresponding entries in the x and w columns. Similarly the s column is the cross-product of the x and z columns. The consequence of proceeding in this way will be discussed in the next section when the design is completed.

It would, of course, be absolutely impossible to estimate quadratic effects from the results of the experiment described by Table 14.2, but we still have two trials at our disposal and these will be given an intermediate level for each of the five independent variables (Table 14.3).

Having completed the design matrix we can translate it into a working design

Table 14.3   Additional trials

			Variable		
Trial	x	w	z	t	s
9	0	0	0	0	0
10	0	0	0	0	0

Table 14.4   Allocation of real values to the independent variables

Artificial value	Weight of special ingredient	Feedrate	Inlet temperature	Agitation speed	Catalyst age
	x	z	t	s	w
−1	0	1	1	2	1, 2, 4, 5
0	3	4	2	3	3, 8
+1	6	7	3	4	6, 7, 9, 10

if we replace −1, 0 and +1 by realistic values of the five variables. Staying within the limits used by the plant manager we will translate as in Table 14.4

Catalyst age (*w*) must be allocated values 1 to 10 and it might seem logical to replace −1 with 1, 2, 3, 4 and 0 with 5, 6 etc. On the other hand perhaps it is desirable to separate the two identical trials which have the intermediate value, '0', so these are allocated 3 and 8. Inserting the revised values of the independent variables in the design matrix gives the design of Table 14.5.

Because of the peculiar nature of one of the independent variables a further change is needed before the experiment can be carried out. In practice, catalyst

Table 14.5

Trial	Weight of special ingredient	Catalyst age	Feedrate	Inlet temperature	Agitation speed
	x	w	z	t	s
1	0	1	1	3	4
2	6	2	1	1	2
3	0	6	1	1	4
4	6	7	1	3	2
5	0	4	7	3	2
6	6	5	7	1	4
7	0	9	7	1	2
8	6	10	7	3	4
9	3	3	4	2	3
10	3	8	4	2	3

age (w) must increase in steps of 1 throughout the experiment if we are using consecutive batches. The trials in Table 14.5 will, therefore, need to be re-ordered to conform to this constraint. Reordering gives the final design in Table 14.6.

Table 14.6   The final design

Trial	Weight of special ingredient x	Catalyst age w	Feedrate z	Inlet temperature t	Agitation speed s
1	0	1	1	3	4
2	6	2	1	1	2
3	3	3	4	2	3
4	0	4	7	3	2
5	6	5	7	1	4
6	0	6	1	1	4
7	6	7	1	3	2
8	3	8	4	2	3
9	0	9	7	1	2
10	6	10	7	3	4

This design has been produced without any consideration of the difficulties of implementation. You will note, for example, that each batch requires a different weight of special ingredient from the preceding batch and that only occasionally do any of the other independent variables keep the same value in two consecutive batches. This constitutes a radical departure from routine production in which the plant operators are encouraged to repeat the *same* procedure as closely as possible. To leave the experiment in the hands of the operators might, therefore, be very unwise. Excessive intervention by management, on the other hand, may result in disturbance of the many other variables which have *not* been included in the experiment, thus making the ten batches non-typical and invalidating any conclusions. Clearly, the use of statistics cannot overcome these difficulties but it can help us to assess the quality of the experiment *before* it is carried out. This we will now attempt.

14.3   How good is the alternative experiment?

We noted in an earlier chapter that it is possible to detect the shortcomings of an experimental design by checking for intercorrelation of the independent variables. The correlation matrix for the design given in Table 14.6 is illustrated in Table 14.7.

Apart from the small positive correlation (0.16) between x and w this correlation matrix is perfect. We may be tempted to conclude, therefore, that this experiment would have given unambiguous estimates of the linear main effects of each of the five independent variables. And so it would *provided that*

Table 14.7   Correlation matrix for the design in Table 14.6

	x	w	z	t	s
x	1.00	0.16	0.00	0.00	0.00
w		1.00	0.00	0.00	0.00
z			1.00	0.00	0.00
t				1.00	0.00
s					1.00

all interactions are non-existent. It is unfortunately true that, if we were to extend the matrix to include columns for cross-product variables, (*xz*, *xw*, etc.), we would find several entries of 1.00 other than those on the diagonal.

These unwanted correlations could have been predicted when we decided to use Table 14.2 as the starting point for our design. By the way in which the *t* and *s* columns were generated we could see that:

Main effect *t* would be confounded with interaction *xw*;
Main effect *s* would be confounded with interaction *xz*.

So we would expect the correlation between *t* and *xw* to be 1.0 and the same could be said of the correlation between *s* and *xz*. There would be several other correlations of +1.0 and these are indicated by the alias groups in Table 14.8.

Table 14.8   Alias groups of the design in Table 14.2

$$
\begin{array}{llll}
(x, & tw, & sz, & xwzts) \\
(z, & sx & wts, & xzwt) \\
(w, & tx, & zts, & xzws) \\
(t, & xw, & wzs, & xzts) \\
(s, & xz, & wzt, & xwts) \\
(zw, & xzt, & xws, & xts) \\
(zt, & xwz, & xst, & xws)
\end{array}
$$

An extended correlation matrix would show a correlation of +1.0 between any pair of variables in the same alias group. Thus main effect *x* would be perfectly correlated with the interaction variables *tw*, *sz* and *xwzts*. Any estimate of main effect *x* would only be valid, therefore, if the interactions did not exist.

In summary, then, we can see that this alternative design has certain shortcomings but it is an improvement on the experiment carried out by the plant manager. Furthermore it is fair to say that *any* experiment which attempts to investigate five independent variables in only ten trials will have similar imperfections. If the plant manager were to abandon the results of his first experiment and to start again he could not hope to achieve very much if he were only prepared to include ten batches. Perhaps he would achieve much more by extending his original experiment in such a way as to remove its major defects.

14.4 Extending the original experiment

You will recall that we first expressed reservations about the plant manager's experiment when we examined a correlation matrix in Chapter 11. That matrix is reproduced in Table 14.9.

Table 14.9 Correlation matrix of the original experiment

	x	w	z	t	s
x	1.00	−0.26	0.97	−0.07	0.00
w		1.00	−0.20	−0.05	0.44
z			1.00	0.07	−0.06
t				1.00	0.00
s					1.00

The major defect in the experiment is indicated by the correlation of 0.97 between x and z. Any extension of the experiment would need to contain trials in which the values of x and z were carefully selected so as to reduce this correlation. How this might be achieved is indicated in Fig. 14.1 which shows the original ten trials plus a further eight trials which could be used to obtain a better overall design.

By introducing the eight additional points into Fig. 14.1 we reduce the correlation between x and z from +0.97 to 0.05. If we now find suitable values for t, s and w to go with these x and z values we then have an additional experiment of eight trials which will form a very useful extension to the original investigation. Note that the eight additional trials by themselves do *not* constitute a good experiment but the whole series of 18 trials does.

In the original experiment there was zero correlation between s and t. In fact we can see in Fig. 14.2 that the values of agitation speed and inlet temperature

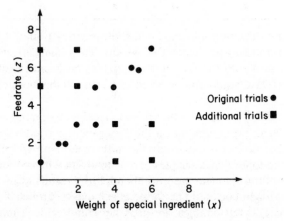

Figure 14.1 Reducing the correlation between x and z

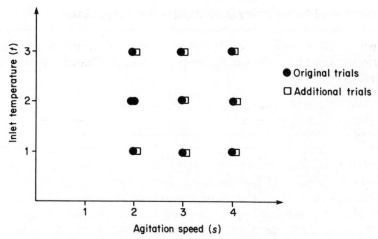

Figure 14.2   Maintaining the zero correlation between *s* and *t*

must have been carefully planned. The eight additional points in Fig. 14.2 make use of the same values and the whole grid of 18 points constitutes two replicates of a 3^2 factorial experiment.

Catalyst age (w) is again a special case. If we start the series of eight batches with a new catalyst then w will vary from 1 to 8. When allocating these values to the trials we will attempt to reduce the correlation of 0.44 which exists between w and s in the original experiment. Combining at random the values of the independent variables from Fig. 14.1 and Fig. 14.2, then carefully adding values for catalyst age we get the experiment listed in Table 14.10.

Table 14.10   An extension to the original experiment

Trial	Weight of special ingredient x	Catalyst age w	Feedrate z	Inlet temperature t	Agitation speed s
11	4	1	3	2	4
12	0	2	7	1	4
13	4	3	1	3	4
14	2	4	5	3	2
15	2	5	7	1	2
16	0	6	5	2	3
17	6	7	1	3	3
18	6	8	3	1	3

The effectiveness of this extension to the original experiment can be assessed by inspection of the correlation matrix in Table 14.11. (Note that this matrix is based on *all 18 trials*.) Clearly the large correlations of Table 14.9 have been greatly reduced by the extension and we have managed to avoid any increase in the correlations which were already satisfactory in the original experiment.

Table 14.11   Correlation matrix for all 18 trials

	x	w	z	t	s
x	1.00	0.02	0.05	0.10	0.03
w		1.00	−0.19	−0.05	0.05
z			1.00	−0.29	−0.20
t				1.00	0.00
s					1.00

It is safe to conclude from the very small intercorrelations of the independent variables in Table 14.11 that the major defect of the original experiment has been eliminated. Can we, therefore, expect to be able to estimate all five main effects after the additional eight trials have been carried out? We can be sure that no pair of main effects will form an alias pair but there is still a possibility that:

(a) either a main effect will be aliased with an interaction
(b) or two interactions will form an alias pair.

To check these possibilities we will extend the correlation matrix by introducing cross-product variables. Before doing so we will, of course, scale the independent variables. The regression package used for this analysis subtracts the mean then divides by the standard deviation:

$$X = (x - \bar{x})/s_x \qquad W = (w - \bar{w})/s_w \qquad \text{etc.}$$

Table 14.12   The extended experiment

	X	W	Z	T	S	XW	XZ	XT	XS	WZ	WT	WS	ZT	ZS	TS	XX	WW	ZZ	TT	SS
X		02	05	10	03	30	00	08	20	27	35	01	22	11	16	00	20	00	17	40
W			19	05	05	20	21	32	01	05	28	68	07	03	09	33	12	03	03	15
Z				29	20	26	00	23	13	03	09	03	19	15	20	00	04	00	06	00
T					00	34	21	12	19	09	02	09	04	20	00	09	28	25	00	00
S						01	10	16	32	03	10	10	19	00	00	21	71	19	00	00
XW							56	10	05	18	09	03	17	47	24	19	20	00	10	04
XZ								39	05	03	16	39	40	11	27	04	09	04	19	15
XT									08	17	10	23	06	32	29	08	02	45	07	14
XS										59	28	01	35	20	00	41	00	25	53	03
WZ											15	06	10	10	24	54	32	34	34	16
WT												17	43	22	12	18	14	27	04	09
WS													22	08	11	01	31	13	09	04
ZT														28	51	16	18	63	20	33
ZS															53	29	25	15	48	13
TS																09	16	45	00	00
XX																	14	43	08	38
WW																		23	20	01
ZZ																			37	13
TT																				00

This differs from what we did in earlier chapters when we divided by half of the range.

The complete correlation matrix, containing five quadratic variables in addition to the ten interactions, is rather large and so it has been printed without decimal points and with negative signs placed under the numbers, as Table 14.12.

The largest correlation coefficients in Table 14.12 are:

    0.71 between S and WW
    0.68 between W and WS
 -0.63 between ZT and ZZ
    etc.

Clearly the design is not perfect but it is doubtful if we could improve upon it to any great extent without increasing the number of trials. Perhaps the next step forward should be to carry out the eight additional trials then make use of the regression package to analyse the whole set of results. We can refer back to Table 14.12 to check particular correlation coefficients when we see which variables are included in the regression equation.

14.5 Final analysis

The plant manager has carried out the eight additional trials using eight consecutive batches starting with a new catalyst. The impurity determinations made on the eight batches are given in the full set of results in Table 14.13.

The data in Table 14.13 are fed into the regression package. The five independent variables are scaled [e.g. $X = (x-3)/2.055$] then the quadratic variables (i.e. X^2, W^2 etc.) and the interaction variables (i.e. XW, XZ, etc.) are generated. This gives us a total of 20 independent variables and all of these are contenders for inclusion in the regression equation. The equation builds up as follows:

$y = 4.56 - 1.62X$ 61.97% fit
$y = 4.56 - 1.71X + 0.919T$ 81.64% fit
$y = 4.52 - 1.70X + 0.775T + 0.596XZ$ 92.02% fit
$y = 4.56 - 1.68X + 0.835T + 0.485XZ - 0.291XT$ 93.68% fit
$y = 4.55 - 1.69X + 0.827T + 0.406XZ - 0.264XT$
      $+ 0.260WS$ 95.01% fit
$y = 4.84 - 1.77X + 0.831T + 0.431XZ - 0.277XT$
      $+ 0.254WS - 0.289SS$ 95.81% fit
    etc.   etc.

To test the significance of each variable we will use the t-test which has served us well in this capacity in previous chapters. The test statistic at each stage has been calculated and is given in Table 14.14.

Table 14.13 The full set of data for the final analysis

Batch number	Impurity	Weight of special ingredient	Catalyst age	Main ingredient		Agitation speed
				Feedrate	Inlet temperature	
	y	x	w	z	t	s
1	4	3	1	3	1	3
2	3	4	2	5	2	3
3	4	6	3	7	3	3
4	6	3	4	5	3	2
5	7	1	5	2	2	2
6	2	5	6	6	1	2
7	6	1	7	2	2	2
8	10	0	8	1	3	4
9	5	2	9	3	1	4
10	3	5	10	6	2	4
11	3	4	1	3	2	4
12	4	0	2	7	1	4
13	4	4	3	1	3	4
14	6	2	4	5	3	2
15	4	2	5	7	1	2
16	7	0	6	5	2	3
17	2	6	7	1	3	3
18	2	6	8	3	1	3
Mean	4.56	3.00	5.06	4.00	2.00	3.00
SD	2.061	2.055	2.676	2.082	0.816	0.816

Table 14.14 Significance of the independent variables

Variable entering the equation	% fit	Test statistic	Degrees of freedom	Critical values	
				5%	1%
X	61.97	5.11	16	2.12	2.92
T	81.64	4.01	15	2.13	2.95
XZ	92.02	4.26	14	2.15	2.98
XT	93.68	1.85	13	2.16	3.01
WS	95.01	1.79	12	2.18	3.05
SS	95.81	1.45	11	2.20	3.11
etc.	etc.	etc.	etc.	etc.	etc.

Had we allowed the computer program to decide which variables were to be included in the equation and specified a 5% significance level, then the program would have stopped after X, T and XZ had been included. The 'best' regression equation would have been printed out as:

$$Y = 4.52 - 1.70X + 0.775T + 0.596XZ$$

There are reasons why we should not accept, without question, the recommendation of the regression package. These include:

(a) When we descale the variables using $x = 3.0 + 2.055X$, $t = 2.0 + 0.816T$ and $z = 4.0 + 2.082Z$ we will find that descaling the cross-product variable XZ has introduced z into the equation. For this reason many statisticians would advise that whenever a cross-product variable is entered into a regression equation, the two linear variables should also be included. Following this advice would lead us to fit an equation with X, T, XZ and also Z as independent variables.

(b) The fourth variable to enter the equation in Table 14.14 was XT. Though its entry was not statistically significant the interaction between X and T is worth pursuing because both X and T were already in the equation. Further analysis of the data might therefore include an equation with X, T and XT as independent variables. We see in Table 14.12 that the correlation between XZ and XT is -0.39, which is not negligible, so the inclusion of XT as the third variable *might* give a significant increase in percentage fit.

With these thoughts in mind we run the computer program again to obtain the following equations:

$$y = 4.52 - 1.68X + 0.699T + 0.610XZ - 0.241Z \qquad 93.25\% \text{ fit}$$
$$y = 4.61 - 1.68X + 0.978T - 0.517XT \qquad 88.04\% \text{ fit}$$

In the first of these equations we see that the introduction of Z as the fourth independent variable increases the percentage fit from 92.02% to 93.25%. This increase is not statistically significant but the purpose of introducing Z was simply to 'balance' the equation.

The second of the two equations has a surprisingly high percentage fit and the introduction of XT as a third variable has resulted in an increase from 81.64% to 88.04%. This increase *is* statistically significant with a test statistic equal to 2.74.

In the light of what has been revealed by these two equations we might return to the computer for further use of the regression package. One possibility is to fit an equation which includes X, T, Z, XZ and XT as independent variables. We might even try ZT as a sixth variable. All of this analysis can be carried out very quickly on our desk top computer but we must call a halt at some point and attempt to translate our findings into language that will be understood by plant personnel. Further improvements in the equation are of doubtful value as we already have sufficient evidence on which to base confident conclusions.

For the sake of simplicity we *might* wish to base conclusions on the second regression equation ($y = 4.56 - 1.71X + 0.919T$) which accounts for 81.64% of the batch to batch variation in impurity. Descaling X and T gives us:

$$y = 4.81 - 0.83x + 1.13t$$

We can expect therefore a reduction in impurity of 0.83% to result from a unit increase in the weight of special ingredient (x). We could expect a further increase in impurity of 1.13% for a unit decrease in the inlet temperature (t) of the main ingredient. *Note* that we can expect *both* of these benefits, not just one or the other, since there is very little correlation between x and t.

The absence of z in the above equation resolves the dilemma that remained unresolved in Chapter 11. You will recall that we were unable to decide whether the pigment impurity was dependent on the weight of special ingredient (x) *or* on the feedrate (z). *We can now conclude that x is important whereas z can be ignored.* This statement must be qualified with the reservation that changing the value of feedrate (z) *might* have some effect on pigment impurity if it were set to a value outside the range used in the experiment. It is always possible that

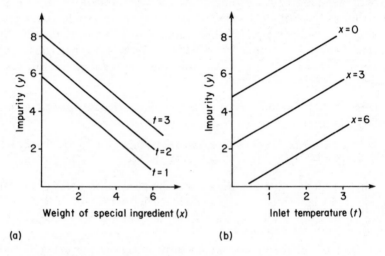

Figure 14.3   Graphical representation of $y = 4.81 - 0.83x + 1.13t$

we will fail to detect the effect of an independent variable if its range of values in the investigation is too small.

Many regression packages would print out confidence intervals for the coefficients in the equation. If we had a confidence interval for the coefficient of x we would predict a decrease in impurity within a certain range rather than quoting one figure, 0.83%. Such confidence intervals can be useful but far more important to the plant manager is a graphical representation which will help him to choose values of x and t which can be expected to give a tolerable level of impurity. Fig. 14.3 may be useful in this respect.

From Fig. 14.3(a) or (b) we can see that a low inlet temperature and a high weight of special ingredient are required in order to obtain a low level of impurity. There are, of course, many combinations of the two independent variables which will give an expected impurity of less than 2%, say. This point may be even more clear in Fig. 14.4, which is a *contour diagram* representing the same equation as Fig. 14.3.

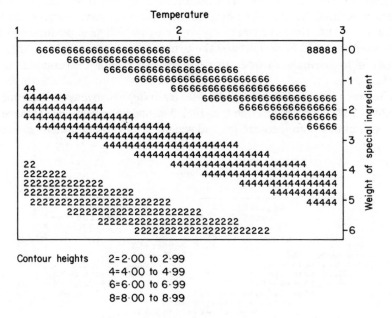

Temperature

Contour heights   2 = 2·00 to 2·99
4 = 4·00 to 4·99
6 = 6·00 to 6·99
8 = 8·00 to 8·99

Figure 14.4   Contour diagram of $y = 44.81 - 0.83x + 1.13t$

A contour diagram is even more useful if the regression equation contains quadratic or cross-product terms. The alternative equation which contained the XT interaction (i.e. $y = 4.61 - 1.68X + 0.978T - 0.517XT$) after descaling becomes:

$$y = 2.82 - 0.192x + 2.12t - 0.308xt$$

This equation is represented by the contour diagram in Fig. 14.5.

A comparison of Fig. 14.4 and Fig. 14.5 reveals strong similarities overall. Nonetheless, the small differences between the two contour diagrams may be of practical importance. We will therefore concentrate on Fig. 14.5.

The numbers (2, 4, 6, 8) in the contour diagram represent contour bands or ranges of impurity. Thus a 4 indicates that the predicted impurity level lies between 4.00% and 4.99% for that particular combination of temperature (t) and weight of special ingredient (x). For minimum impurity we need to be in the bottom left hand corner of Fig. 14.5. In practical terms we need to use a low inlet temperature and a high weight of special ingredient. This conclusion could easily have been drawn from an inspection of the regression equation; so why do we need a contour diagram? A contour diagram would have been more useful to us if:

(a) the regression equation had been more complex;
(b) we had wished to optimize two or more dependent variables.

Confronted with a contour diagram for impurity and a second for yield we might well find that the operating conditions which give minimum impurity differ somewhat from the operating conditions which give maximum yield. A compromise could be sought and the contour diagrams would be particularly useful in this search for operating conditions which are acceptable on both counts.

There is a third independent variable (feedrate) which may be important but which is completely ignored in Fig. 14.5. We had an indication that there is an interaction between feedrate (z) and weight of special ingredient (x) but xz is

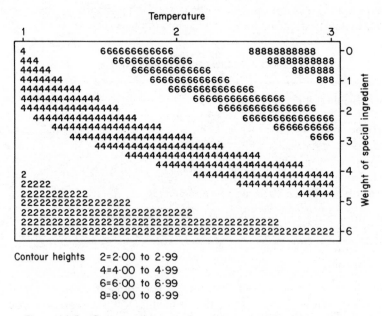

Figure 14.5   Contour diagram of $y > 2.82 - 0.192x + 2.12t - 0.308xt$

not included in the equation which gave Fig. 14.5. To accommodate this interaction and also the xt interaction we could fit an equation involving X, Z, T, XZ and XT. After descaling of the variables this equation could be represented by a *series* of contour diagrams similar to Fig. 14.5 but each having a different value of feedrate (z). Alternatively we could produce a series of contour diagrams using x and z as the independent variables with each diagram having a different value of temperature (t). Such a set of diagrams would be a useful aid in selecting suitable operating conditions for future batches.

This discussion has embraced many *possibilities*, but has said little about the immediate needs of the plant manager. He is paid a handsome salary for *making decisions* concerning the safety and efficient operation of the plant, and he must now specify operating conditions for future batches. These will include values for the five independent variables in the investigation and his specification is:

> Weight of special ingredient   5   units
> Feedrate of main ingredient   4   units
> Temperature of main ingredient   1.5   units
> Agitation speed   3   units
>
> Regulations concerning catalyst changes are to continue as before.

The plant manager is not entirely happy but he hopes that future experiments will add further to his growing understanding of the plant.

14.6   Inspection of residuals

In the preceding analysis we have concentrated on reaching decisions about *which* of the independent variables are important and what values they need to have if the impurity of future batches is to be reduced. Multiple regression proved to be a very useful tool in this analysis.

We have seen in earlier chapters some of the dangers inherent in this very powerful technique but we cannot claim to have given thorough coverage to all the *assumptions* which underlie regression analysis. Restrictions on the size of this book prevent such coverage and the reader is referred to one of the many excellent texts in the Bibliography for completion of his/her regression education. It would, however, be irresponsible to close without pointing to one simple way of checking that assumptions are not being violated. This method involves examination of the *residuals* from the regression equation, and such examination should be carried out before using the equation as a basis for drawing conclusions. This examination is facilitated if the regression package fitting the equation also prints out a graphical representation of the residuals.

It is highly desirable that the residuals should have a distribution which approximates to the normal distribution. Whilst a normal distribution of residuals does not guarantee that *all* assumptions are satisfied, any clear departure from normality is a strong indication that something is amiss. For this reason some regression packages print out a probability plot of the residuals.

It is highly desirable that the residuals should not contain a serial pattern when plotted against 'time' or against any of the independent variables. To aid detection of such patterns some regression packages print out a cusum plot of the residuals. Such a graph can be very revealing.

Fig. 14.6 is a cusum plot of the residuals from one of the equations examined earlier ($y = 4.61 - 1.68X + 0.978T - 0.517XT$). Clearly there is a pattern here which cannot be ignored. Furthermore, the maximum cusum is found to be statistically significant, indicating a change in the mean level of the residuals at batch number 11.

Table 14.15   Residuals from the regression equation

Batch number	Residual	Batch number	Residual	Batch number	Residual
1	0.59	7	−0.24	13	−0.93
2	−0.79	8	0.82	14	0.12
3	1.57	9	1.08	15	−0.93
4	0.20	10	0.03	16	−0.06
5	0.76	11	0.08	17	−0.79
6	−0.39	12	−0.43	18	−0.68

Even if the regression package did not produce a cusum plot of the residuals we would almost certainly have spotted the change simply by glancing at a *list* of the residuals. The pattern of + and − signs is very striking in Table 14.15 which contains the residuals represented in Fig. 14.6.

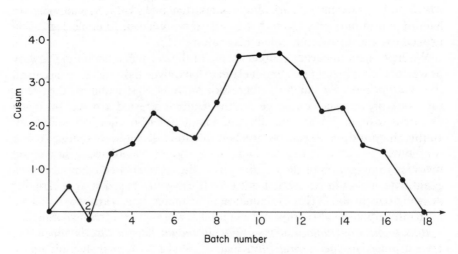

Figure 14.6   Cusum plot of residuals from regression equation

Most of the early batches have positive residuals whilst a majority of the later batches have negative residuals. The mean of *all* 18 residuals is, of course, zero and there appears to be a change around batch number 11 from positive to negative. An explanation for this change springs to mind immediately. The whole experiment of 18 trials was actually carried out in two halves or two blocks with an interval of many days between batch 10 and batch 11. The plant manager made every effort to repeat the circumstances surrounding the first ten batches when manufacturing the last eight but the cusum plot suggests that he was not entirely successful.

The change in the mean level of the residuals is statistically significant, but is it important? The mean of the residuals for batches 1 to 10 is +0.363 and the mean for batches 11 to 18 is −0.453 which indicates a decrease in impurity of 0.816% in the later batches. Certainly a decrease of almost 1% cannot be ignored especially as this is a change over and above the variation in impurity accounted for by the independent variables in the regression equation.

So the change in the residuals is indicative of a change in impurity which is both significant *and* important, but does this new finding invalidate our early conclusions? Probably not, because both of the prime variables, x and t, have similar mean values in both halves of the experiment. We can, however, resolve the uncertainty by introducing an extra independent variable which distinguishes between the early batches and those produced later. This variable would be equal to 0 for trials 1 to 10 and equal to 1 for trials 11 to 18. Submitting this extended data set to the regression package would probably result in X being the first independent variable and T being the second (as we found earlier); then it is possible that the new variable would enter followed by an interaction, probably XZ or XT.

Perhaps the inspection of residuals has emphasized, yet again, that we cannot expect to analyse a set of data in one run of a regression package. The researcher would be well advised to regard data analysis as an interactive process in which computer runs are interspersed with periods of contemplation in which results and graphical representations are scrutinized.

14.7 Summary

In this chapter we have completed the circle by returning to the problem which arose in Chapter 9, and we have, at last, advised the plant manager. In order to draw unambiguous conclusions we had first to extend the rather bad experiment which had already been carried out. The additional trials were carefully selected in order to eliminate the deficiencies that had been identified.

A multiple regression package was used in the analysis of the experimental results and several equations were fitted before we identified inlet temperature (t) and weight of special ingredient (x) as the most important variables to control. We also found that two interactions were quite important.

The use of contour diagrams was helpful when translating the regression equation into a practical strategy for future production. We noted, however, that this final step should not be taken until the residuals from the regression equation have been examined. Such an examination can serve as a check on some of the assumptions underlying regression analysis and, by means of a cusum plot, we spotted a change in the residuals which indicated a need for further analysis. To achieve further progress we could introduce a 'blocking variable' which was equal to 0 for the first block of trials and equal to 1 for the second block of trials.

In the final chapter we will examine a statistical technique which could have been used throughout the second part of this book. We managed very well without this technique but its study will certainly enlarge the reader's understanding of both design and analysis.

Problems

(1) Knaresborough Breweries, well known for their 'light headed' beers, have decided to launch a new 'Strong-ail' beer. Ten experiments have been carried out to determine the effect of three variables on the quality of 'Strong-ail'. Only *main effects* are being investigated with no consideration being given to interactions. The three variables are:

Malt strain	F1 or F4
Fermentation temperature	High or low
Yeast	Enzyme or biozyme

The design of the experiment was as follows:

Cell number	Malt	Temperature	Yeast
1	F1	Low	Bioenzyme
2	F1	High	Enzyme
3	F4	Low	Enzyme
4	F4	Low	Bioenzyme
5	F4	Low	Enzyme
6	F1	High	Enzyme
7	F1	High	Enzyme
8	F4	Low	Bioenzyme
9	F1	Low	Bioenzyme
10	F1	Low	Bioenzyme

(a) List the design matrix in coded form using the values -1 and $+1$. Calculate the correlation between malt and temperature and hence complete the following correlation matrix:

	Malt	Temperature	Yeast
Malt	1.0		0.00
Temperature		1.0	-0.65
Yeast	0.00	-0.65	1.0

Refer to the solution for part (a) before attempting part (b).

(b) Extend the design by choosing two cells which will greatly improve the correlation matrix.

Refer to the solution for part (b) before attempting part (c).

(c) Complete the new correlation matrix given below:

	Malt	Temperature	Yeast
Malt	1.0	-0.17	$+0.17$
Temperature	-0.17	1.0	
Yeast	$+0.17$		1.0

(2) Dr Scratchplan has decided to investigate the effect of temperature, pressure and concentration of catalyst on the whiteness of polymer. He has chosen three levels for each factor.

Temperature (°C)	270, 275, 280
Pressure (p.s.i.)	600, 700, 800
Concentration of catalyst (%)	0.15, 0.20, 0.25

Since he is away for the next week, suffering a course on experimental design, he produces the following plan for his laboratory assistant to carry out.

Stage 1: Set temperature at 275 °C and pressure at 700 p.s.i. and investigate the three levels of concentration of catalyst.
Stage 2: Choosing the concentration of catalyst which gave the whitest polymer, investigate the two remaining levels of pressure keeping temperature at 275 °C.
Stage 3: Choosing the 'best' pressure and concentration of catalyst, investigate the two remaining levels of temperature.

While on the course he realizes that temperature may have a curved relationship with whiteness and also that the interaction (pressure × concentration of catalyst) may be important. He phones his laboratory assistant and is given the levels for the seven cells in the experiment.

Cell	Temperature	Pressure	Concentration
1	275	700	0.15
2	275	700	0.20
3	275	700	0.25
4	275	600	0.25
5	275	800	0.25
6	270	600	0.25
7	280	600	0.25

(a) List the design matrix for temperature, pressure, concentration, $(temperature)^2$ and $(pressure × concentration)$ in coded form using $-1, 0$ and $+1$ for the three levels of each variable.

Refer to the solution for part (a) before attempting part (b).

(b) Complete the following correlation matrix:

	Temp.	Press.	Conc.	$(Temp.)^2$	$(Press. × conc.)$
Temp.	1.0	0.0	0.0	0.0	0.0
Press.		1.0	−0.24	−0.65	
Conc.			1.0	+0.37	−0.24
$(Temp.)^2$				1.0	
$(Press. × conc.)$					1.0

Refer to the solution for part (b) before attempting part (c).

(c) Dr Scratchplan is somewhat concerned about the quality of his experimental design and wishes to extend the experiment but is informed by the laboratory assistant that there is only enough feedstock for a further two cells. Which two should Dr Scratchplan choose?

Refer to the solution for part (c) before attempting part (d)

(d) Evaluate the design by completing the following correlation matrix:

	Temp.	Press.	Conc.	$(Temp.)^2$	(Press. × conc.)
Temp.	1.0	−0.19	−0.26	0.27	0.30
Press.		1.0	−0.24	−0.70	
Conc.			1.0	0.08	−0.24
$(Temp.)^2$				1.0	−0.10
(Press. × conc.)					1.0

———15———
Analysis of variance

15.1 Introduction

It is hoped that the reader will have acquired a working knowledge of multiple regression analysis before starting to read this chapter. If such a worthy objective has been achieved then the reader will realize that multiple regression analysis can be understood without any knowledge at all of a particular statistical technique known as 'analysis of variance'. Why then is it thought necessary to introduce this technique in the final chapter? Three reasons can be offered:

(a) Many computer packages print out multiple regression results in analysis of variance 'language'.
(b) The majority of texts use analysis of variance as a foundation for multiple regression analysis.
(c) Presenting analysis of variance at this time offers a different perspective on many of the problems we have considered in earlier chapters.

It is hoped, therefore, that studying this final chapter will help the reader to consolidate some important concepts and to broaden his/her statistical knowledge whilst exploring a technique which has an even wider field of application than multiple regression analysis itself.

15.2 Variation between samples and within samples

In the plant manager's experiment we had *one* impurity determination on each of ten batches. No mention was made of how these determinations were obtained. In fact the analytical procedure for determining the percentage of this particular impurity in digozo blue pigment had previously been the subject of extensive investigation and considerable controversy. It had been suggested by production personnel that variation in sampling and testing of pigment caused the introduction of additional error which in some cases may have prevented the pigment from meeting the specification.

In order to compare the variation of impurity from sample to sample with the variation due to testing error, an investigation was carried out as follows. One batch was selected from the many batches produced under normal operating conditions. Four samples of pigment were extracted from this batch and five tests were carried out on each sample. The 20 impurity determinations are given in Table 15.1.

Table 15.1   Five tests on each of four samples

Sample	A	B	C	D
Impurity determinations	5.1 5.2 5.3 5.3 5.1	5.3 5.5 5.8 5.5 5.4	5.8 5.5 5.8 5.8 5.6	5.2 5.0 5.3 5.3 5.2
Mean	5.20	5.50	5.70	5.20
SD	0.1000	0.1871	0.1414	0.1225

The data in Table 15.1 will form the basis for comparing the variation due to testing error with the variation due to heterogeneity of material. From these data we can estimate:

(a) the testing standard deviation (σ_t), which is a measure of the variation in impurity determinations that we would get if the test method were applied to pigment that was perfectly homogeneous;
(b) the sampling standard deviation (σ_s), which is a measure of the variation in impurity determinations that we would get if there were no testing error but the pigment exhibited its normal degree of heterogeneity;
(c) a confidence interval for the true impurity (μ) of the whole batch of pigment from which the samples were taken.

In order to obtain these three estimates we will use a technique known as one-way *analysis of variance* and the first step in this procedure is to break down the total variation of impurity into two components. To measure the total variation we will use the total sum of squares which is calculated in columns 2, 3 and 4 of Table 15.2. From each measurement we subtract the overall mean (5.40), square the deviations and then sum the squared deviations to obtain the total sum of squares (1.22).

We have been carrying out similar calculations since Chapter 1. Obviously the total sum of squares is just an old friend with a different name. The reader will realize that dividing the total sum of squares by its degrees of freedom (19) and then taking the square root would give us the standard deviation of the 20 impurity measurements. We do not need this standard deviation, however, for standard deviations cannot be subdivided in the way that we will partition the total sum of squares.

The *within sample sum of squares* is calculated in columns 5, 6 and 7 of Table 15.2. Once again we square deviations from a mean but we use the appropriate sample mean rather than the overall mean. The deviations in column 6 tend to be smaller than those in column 3 and the within sample sum of squares (0.32) is certainly less than the total sum of squares (1.22).

The *between samples sum of squares* is calculated in columns 8 and 9 of Table 15.2. For this calculation we square the deviations of the sample means from the overall mean. The sum of the squares is equal to 0.90. The calculations which are set out in Table 15.2 are also presented in a graphical form in Figs 15.1, 15.2 and 15.3. In all three diagrams the thick vertical lines represent the deviations which are squared and summed.

A third method of calculation, and one which would be preferred in practice, involves the use of the formulae below:

Total sum of squares
$$= \text{(overall SD)}^2 \text{ (degrees of freedom)}$$
Within samples sum of squares
$$= \Sigma\{\text{(sample SD)}^2 \text{ (degrees of freedom)}\}$$
Between samples sum of squares
$$= \text{(SD of sample means)}^2 \text{ (number of samples} - 1)$$
$$\times \text{(number of observations on each sample)}$$

Use of these formulae involves the calculation of six standard deviations but this can be done very quickly with a modern calculator. The 'overall SD' is simply the standard deviation of all 20 measurements, and is equal to 0.2534. The four 'sample SDs' are given in Table 15.1 and the 'SD of sample means' is obtained by entering the four sample means into the calculator. Using these standard deviations we get the following results:

Total sum of squares
$$= (0.2534)^2 \, (20 - 1)$$
$$= 1.22$$
Within samples sum of squares
$$= (0.1000)^2 \, (5 - 1) + (0.1871)^2 \, (5 - 1)$$
$$+ (0.1414)^2 \, (5 - 1) + (0.1225)^2 \, (5 - 1)$$
$$= 0.32$$
Between samples sum of squares
$$= (0.2449)^2 \, (4 - 1) \, (5)$$
$$= 0.90$$

The reader will notice that the use of these formulae has given identical results to those in Table 15.2. Two points should be noted, however:

(a) It is important that the standard deviations are not rounded prematurely. Four significant figures have been used but it would be better to carry *all* the figures given by the calculator.
(b) The third of the three equations must only be used when the same number of tests has been made on each sample, but the first two equations can be used with unequal numbers of observations.

Point (b) need never be a restriction in practice for we only ever need to use *two* of the equations. This is because the 'within samples sum of squares' plus the 'between samples sum of squares' will always equal the total sum of squares.

Table 15.2 Calculation of sums of squares

Sample	Measured impurity	Deviation of impurity from overall mean	Squared deviation	Sample mean	Deviation of impurity from sample mean	Squared deviation	Deviation of sample mean from overall mean	Squared deviation
1	2	3	4	5	6	7	8	9
A	5.1	-0.3	0.09		-0.1	0.01	-0.2	0.04
	5.2	-0.2	0.04		0.0	0.00	-0.2	0.04
	5.3	-0.1	0.01	5.20	0.1	0.01	-0.2	0.04
	5.3	-0.1	0.01		0.1	0.01	-0.2	0.04
	5.1	-0.3	0.09		-0.1	0.01	-0.2	0.04
B	5.3	-0.1	0.01		-0.2	0.04	0.1	0.01
	5.5	0.1	0.01		0.0	0.00	0.1	0.01
	5.8	0.4	0.16	5.50	0.3	0.09	0.1	0.01
	5.5	0.1	0.01		0.0	0.00	0.1	0.01
	5.4	0.0	0.00		-0.1	0.01	0.1	0.01

	Value		Total sum of squares	Mean		Within sample sum of squares		Between samples sum of squares
C	5.8	0.4	0.16		0.1	0.01	0.3	0.09
	5.5	0.1	0.01		−0.2	0.04	0.3	0.09
	5.8	0.4	0.16	5.70	0.1	0.01	0.3	0.09
	5.8	0.4	0.16		0.1	0.01	0.3	0.09
	5.6	0.2	0.04		−0.1	0.01	0.3	0.09
D	5.2	−0.2	0.04		0.0	0.00	−0.2	0.04
	5.0	−0.4	0.16		−0.2	0.04	−0.2	0.04
	5.3	−0.1	0.01	5.20	0.1	0.01	−0.2	0.04
	5.3	−0.1	0.01		0.1	0.01	−0.2	0.04
	5.2	−0.2	0.04		0.0	0.00	−0.2	0.04
Total	108.0	0.0	1.22	—	0.0	0.32	0.0	0.90
Overall mean	5.40							

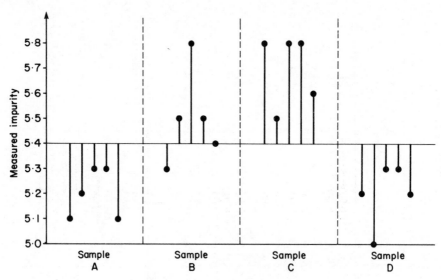

Figure 15.1   Total sum of squares

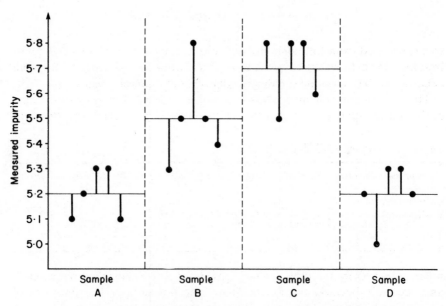

Figure 15.2   Within samples sum of squares

This is easily seen in Table 15.3 where all three sums of squares have been gathered together.

We see also in Table 15.3 that the degrees of freedom can be added in the same manner as the sums of squares. The total sum of squares has 19 degrees of freedom, of course, because it is based on the standard deviation of 20 measurements. The within samples sum of squares has 16 degrees of freedom because it

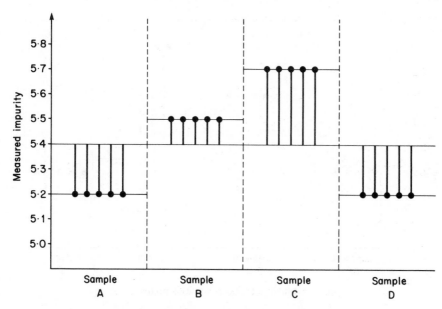

Figure 15.3   Between samples sum of squares

was calculated from four standard deviations, each of which had 4 degrees of freedom. The between samples sum of squares has 3 degrees of freedom because it is based on the standard deviation of four sample means.

The mean squares in an analysis of variance table are calculated by dividing each sum of squares by its degrees of freedom. It is usual to omit the total

Table 15.3   Analysis of variance table

Source of variation	Sum of squares	Degrees of freedom	Mean square
Between samples	0.90	3	0.30
Within samples	0.32	16	0.02
Total	1.22	19	—

mean square from the table for two reasons – firstly because it is not needed, and secondly because the mean squares are not additive like the sums of squares and degrees of freedom. From the mean squares in Table 15.3 we can obtain estimates of the sampling standard deviation (σ_s) and the testing standard deviation (σ_t). The latter is easier to estimate so we will turn to that one first.

The within samples sum of squares is not due to heterogeneity of material but is entirely the result of testing error. If there were no testing error then the data from our experiment would resemble the hypothetical data in Fig. 15.4 and the within samples sum of squares would be equal to zero. It follows, therefore,

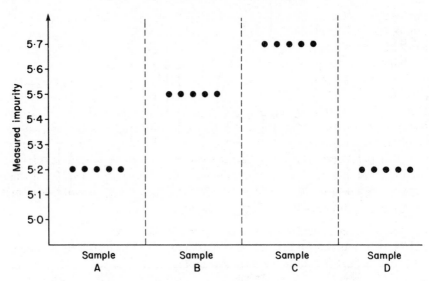

Figure 15.4   Hypothetical data with no testing error

that we can use the square root of the within samples mean square as an estimate of the testing standard deviation.

$$\text{Estimate of testing standard deviation } (\sigma_t) = \sqrt{0.02}$$
$$= 0.1414$$

Estimating the sampling standard deviation is not quite so easy because the between samples sum of squares is influenced by *both* testing error and heterogeneity of the pigment. It is obvious that variation in material would contribute to the between samples sum of squares for we used deviations of the sample means from the overall mean in its calculation. If all four sample means were equal, the between samples sum of squares would be zero. But we would not expect identical sample means even if the pigment were perfectly homogeneous, for the random testing error in each measurement would work its way through to the sample means. For this reason we take into account both the within samples mean square and the between samples mean square when we estimate the sampling standard deviation.

Estimate of sampling standard deviation (σ_s)
$$= \sqrt{[(\text{between samples mean square} - \text{within samples mean square})/}$$
$$(\text{number of measurements on each sample})]$$
$$= \sqrt{\left(\frac{0.30 - 0.02}{5}\right)}$$
$$= 0.237$$

Before we make use of this estimate it would be wise to check the possibility that one of its two contributing influences might not exist. We have suggested that the between samples mean square may be partly due to testing error and partly due to heterogeneity of material. We would only get a mean square equal to zero if *both* of these were absent. Is it possible, then, that our between samples mean square (0.30) is the result of only one of these two causes? Perhaps $\sigma_t = 0$ or perhaps $\sigma_s = 0$.

We can rule out, immediately, the possibility that the testing standard deviation is equal to zero because the *within* samples mean square is not equal to zero. The other suggestion ($\sigma_s = 0$) cannot be rejected so easily, of course, because the testing error alone would give a positive value to the between samples mean square.

It can be shown, however, that the two mean squares would be approximately equal if there were no sampling variation. We have found the between samples mean square (0.30) to be considerably larger than the within samples mean square (0.02). Should we conclude that:

(a) the material is heterogeneous (i.e. $\sigma_s \neq 0$), or
(b) an extremely unusual combination of random testing errors has resulted in very large differences between the sample means.

To help us decide whether (a) or (b) is the most reasonable conclusion we can carry out a significance test. The reader may recall from Part One that we use the *F*-test when we wish to compare two sources of variation.

Null hypothesis – Pigment impurity does not vary from sample to sample throughout the batch (i.e. $\sigma_s = 0$).

Alternative hypothesis: $\sigma_s \neq 0$

$$\text{Test statistic} = \frac{\text{between samples mean square}}{\text{within samples mean square}}$$

$$= \frac{0.30}{0.02}$$

$$= 15.0$$

Critical values – From the one-sided *F*-table with 3 and 16 degrees of freedom:

3.24 at the 5% significance level
5.29 at the 1% significance level.

Decision – We reject the null hypothesis with great confidence.

Conclusion – We conclude that there is variation in impurity throughout the batch.

The very large test statistic implies that the heterogeneity of the material gives rise to variation in impurity measurements which is considerably greater than the variation due to testing error.

Having estimated the sampling and testing standard deviations we could now make use of these estimates to calculate a confidence interval for the true impurity of the whole batch. We could also use the sampling and testing standard deviations for a variety of other purposes which are not within the scope of this book. For a fuller discussion of the use of analysis of variance in this type of situation the reader should refer to *Statistics for Analytical Chemists* by Caulcutt and Boddy (to be published by Chapman and Hall, 1983).

15.3   Analysis of variance and simple regression

In Chapter 10 we fitted a simple regression equation ($y = a + bx$) and then we calculated a residual sum of squares. It was pointed out at the time that this calculation could be based upon the distances of the points from the regression line. These distances, known as residuals, would be squared and the squares added to obtain the residual sum of squares. The method is illustrated in Fig. 15.5.

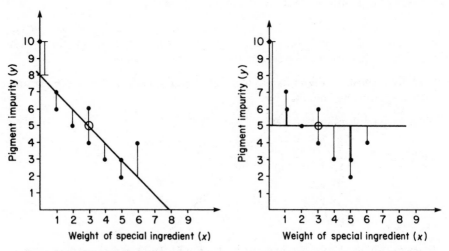

Figure 15.5   Residual sum of squares   *Figure* 15.6   Total sum of squares

Compare Fig. 15.5 with Fig. 15.6. The latter illustrates the calculation of the total sum of squares which is based on deviations from the overall mean ($\bar{y} = 5.0$). Clearly the deviations of the points from the horizontal line in Fig. 15.6 tend to be greater than the deviations from the regression line in Fig. 15.5 and the total sum of squares (50.0) is therefore greater than the residual sum of squares (14.0).

The horizontal line in Fig. 15.6 and the regression line in Fig. 15.5 both pass through the centroid. The method of least squares can be regarded as a process which starts with the horizontal line and rotates it to a new position which minimizes the residual sum of squares. The difference between the total sum of squares and the residual sum of squares is a measure of how much of the variation in impurity has been explained by the regression line. This difference is usually called the 'due to regression sum of squares' or simply 'regression sum of squares' and it could be calculated directly as follows:

$$\text{Regression sum of squares} = r^2(n-1)\,(\text{SD of } y)$$

where r is the correlation between x and y.

You may recall that the percentage fit is equal to $100\ r^2$. It would seem possible therefore to relate the percentage fit, the regression sum of squares and the total sum of squares [which is equal to $(n-1)\,(\text{SD of } y)^2$]. Indeed, there is a very simple relationship:

$$\text{Percentage fit} = \frac{\text{regression sum of squares}}{\text{total sum of squares}} \times 100$$

The regression sum of squares can also be illustrated graphically. In Fig. 15.7 each of the ten points represents a batch but the *predicted* impurity values were used in plotting the points, rather than the actual impurity values used in Figs 15.5 and 15.6.

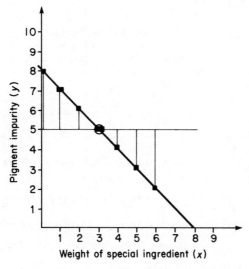

Figure 15.7   Due to regression sum of squares

It is interesting to contrast Figs 15.5, 15.6 and 15.7 on the one hand with Figs 15.1, 15.2 and 15.3 on the other. Comparing Fig. 15.6 with Fig. 15.1 we see that the total sum of squares is calculated in the same way for both sets of data, as it would be calculated for *any* set of data. Comparing Fig. 15.5 with Fig. 15.2 we see a similarity between the residual sum of squares and the within samples sum of squares. In both cases we are measuring the deviations of the points from a fitted line; in Fig. 15.5 it is a regression line, whilst in Fig. 15.2 it is a stepped horizontal line which passes through the sample means. To complete the comparison we see a similarity between the regression sum of squares in Fig. 15.7 and the between samples sum of squares in Fig. 15.3. In both of these diagrams we are measuring the deviation of the fitted line from the horizontal line.

With both sets of data we are using analysis of variance to break down the total variation into two components. In both cases we could refer to one component as 'assignable variation' because it is introduced intentionally, by changing weight of special ingredient in one case and by taking different samples in the other. The second component of variation could be described as 'unassignable variation' or 'residual variation' since it has not been accounted for.

To complete our application of analysis of variance to the simple regression example we can draw up an analysis of variance table and follow this with an *F*-test. In Table 15.4 we see again the additive property of the sums of squares and the degrees of freedom. The residual sum of squares has $(n-2)$ degrees of freedom (as it did in Chapter 10) because we calculated *two* constants, the slope and the intercept, when fitting the equation. The due to regression sum of squares has 1 degree of freedom because we have one independent variable in the equation.

Table 15.4   Analysis of variance table (simple regression)

Source of variation	Sum of squares	Degrees of freedom	Mean square
Due to regression on *x*	36.0	1	36.0
Residual	14.0	8	1.75
Total	50.0	9	—

Null hypothesis – There is no relationship between weight of special ingredient (x) and pigment impurity (y) (i.e. the true slope, β, is equal to zero).

Alternative hypothesis – There is a relationship between x and y (i.e. $\beta \neq 0$).

Test statistic $= \dfrac{\text{due to regression on } x \text{ sum of squares}}{\text{residual sum of squares}}$

$= \dfrac{36.0}{1.75}$

$= 20.57$

Critical values – From the one-sided *F*-table with 1 and 8 degrees of freedom:

5.32 at the 5% significance level
11.26 at the 1% significance level.

Decision – We reject the null hypothesis at the 1% level of significance.

Conclusion – We conclude that, within the population of batches, pigment impurity (y) *is* related to the weight of special ingredient (x).

We have, of course, reached this same conclusion twice already. The first occasion was when we tested the significance of the correlation between x and y, in Chapter 10. The second occasion was also in Chapter 10, when we tested the significance of the slope b (or the significance of the percentage fit, if you prefer that viewpoint). The one-sided F-test above will always lead us to the same conclusion as the two-sided t-test or the two-sided correlation test.

15.4 Analysis of variance with multiple regression

The reader may recall that, in Chapter 11, a second independent variable was introduced into the regression equation and the percentage fit increased from 72% to 93.4%. It was demonstrated by means of a t-test that this increase in percentage fit was statistically significant and we concluded that the second independent variable (t) was related to the pigment impurity. We concluded that it was important to control both the weight of special ingredient (x) *and* the inlet temperature (t) in the production of future batches.

Let us now apply analysis of variance to this second step of the multiple regression analysis. After fitting the second equation we calculated a new set of residuals in Table 11.4. By squaring these residuals and summing the squares we obtain the new residual sum of squares which is equal to 3.30. This figure is, of course, less than the residual sum of squares from the first equation (14.00). Since the total sum of squares remains unchanged by the introduction of another independent variable, any reduction in the residual sum of squares must be matched by an increase in the regression sum of squares as we can see in Table 15.5.

Table 15.5 Analysis of variance with two independent variables

Source of variation	Sum of squares	Degrees of freedom	Mean square
Due to regression on x and t	46.70	2	23.35
Residual	3.30	7	0.471
Total	50.00	9	—

Note that the regression sum of squares now has 2 degrees of freedom because we have two independent variables in the equation. The residual degrees of freedom become $(n - 3)$ because we have calculated three constants from the data (i.e. one intercept and two slopes). We can, as always, follow the analysis of variance table with a one-sided F-test to check the significance of the two independent variables.

Null hypothesis – Pigment impurity is not related to weight of special ingredient (x) and/or inlet temperature (t).

Alternative hypothesis – Pigment impurity not related to weight of special ingredient (x) or inlet temperature (t).

$$\text{Test statistic} = \frac{\text{due to regression on } x \text{ and } t \text{ mean square}}{\text{residual mean square}}$$

$$= \frac{23.35}{0.471}$$

$$= 49.57$$

Critical values – From the one-sided F-table with 2 and 7 degrees of freedom:

4.74 at the 5% significance level
9.55 at the 1% significance level.

Decision – We reject the null hypothesis at the 1% significance level.

Conclusion – We conclude that pigment impurity is related to weight of special ingredient and/or inlet temperature.

This significance test is perfectly valid but it has served no useful purpose. To reach a conclusion that 'y is dependent on x and/or t' does not constitute a step forward as we have already concluded that 'y is dependent on x'. The increase in percentage fit when t is introduced may or may not be significant, but Table 15.5 and the subsequent F-test do not help us to decide. To progress further we must take the regression sum of squares from Table 15.5 and subdivide it into two components. These are the two entries above the broken line in Table 15.6.

Table 15.6   Separating the effects of the two variables

Source of variation	Sum of squares	Degrees of freedom	Mean square
Due to regression on x	36.00	1	—
Due to introduction of t	10.70	1	10.70
Due to regression on x and t	46.70	2	—
Residual	3.30	7	0.471
Total	50.00	9	—

The sums of squares above the broken line add up to the sum of squares immediately below the line. The same is true of the degrees of freedom. The entries below the broken line are taken from Table 15.5 whilst the 'due to regression on x' row is taken from Table 15.4. Obviously, we have broken down the regression sum of squares (46.70) into two components, the first being the 36.00 which was tested earlier and the second being 10.70 which has resulted from the introduction of the second independent variable and which will now be subjected to an F-test.

Null hypothesis – Pigment impurity is not related to inlet temperature (t).

Alternative hypothesis – Pigment impurity is related to inlet temperature (*t*).

$$\text{Test statistic} = \frac{\text{due to inclusion of } t \text{ mean square}}{\text{residual mean square}}$$

$$= \frac{10.70}{0.471}$$

$$= 22.72$$

Critical values – From the one-sided *F*-table with 1 and 7 degrees of freedom:

5.59 at the 5% significance level
12.25 at the 1% significance level.

Decision – We reject the null hypothesis at the 1% significance level.

Conclusion – We conclude that the pigment impurity (*y*) is dependent on the inlet temperature (*t*) as well as the weight of special ingredient (*x*).

Use of analysis of variance and the one-sided *F*-test has led us to exactly the same conclusion that we reached in Chapter 11 when using the two-sided *t*-test. Many multiple regression packages print out analysis of variance tables and carry out *F*-tests. This may be quite acceptable to the experienced user but the beginner would surely find the *t*-test easier to follow. In expressing this opinion the author does not wish to denigrate analysis of variance. There can be no better foundation on which to build an understanding of many interrelated multivariate techniques but analysis of variance can prove a difficult obstacle for the novice to overcome.

15.5 Analysis of variance and factorial experiments

This very versatile technique can be used to analyse the results of a factorial experiment. By means of analysis of variance we can calculate a mean square for each main effect and for each interaction. The calculations are similar to those in Table 15.2 when we carried out a one-way analysis of variance to separate the within samples and the between samples sums of squares. For a 2^2 factorial experiment we would use two-way analysis of variance as follows.

Table 15.7 Results of a 2^2 factorial experiment (from table 12.4)

Feedrate	Temperature (°C)	
	Low (60)	High (70)
Low (40)	76	72
High (50)	70	78

Table 15.8

Temperature	Yield	Deviation of yield from overall mean	Squared deviation	Group mean yield	Deviation of yield from group mean	Squared deviation	Deviation of group mean from overall mean	Squared deviation
1	*2*	*3*	*4*	*5*	*6*	*7*	*8*	*9*
Low	76	2	4	73	3	9	−1	1
	70	−4	16		−3	9	−1	1
High	72	−2	4	75	−3	9	1	1
	78	4	16		3	9	1	1
Total	296	0	40	—	0	36	0	4
Mean	74							

Column 4: Total sum of squares

Column 7: Within temperatures mean square

Column 9: Between temperatures mean square

We will use the results of the 2^2 factorial experiment in Table 15.7 to calculate:

(a) Between feedrates sum of squares.
(b) Between temperatures sum of squares.
(c) Interaction sum of squares.

These sums of squares could equally well be called 'between rows sum of squares', 'between columns sum of squares' and 'between diagonals sum of squares' because of the way in which they are calculated. Each of the sums of squares is based upon the deviation of group means from the overall mean. There are three possible groupings, however, and in Table 15.8 the yields of the four batches are split into two groups depending upon the temperature that was used in the manufacture.

Clearly the calculations in Table 15.8 are based on *exactly the same* procedure as that followed in Table 15.2. We have carried out a one-way analysis of variance using one of the independent variables, temperature, as a criterion for splitting the yield values into two groups. If we had used the other independent variable, feedrate, as a grouping criterion then the group means would have been the two row means of Table 15.8 and we would have obtained:

Total sum of squares = 40
Within feedrate sum of squares = 40
Between feedrates sum of squares = 0

A third method of classification using the diagonal means would have given:

Total sum of squares = 40
Within diagonals sum of squares = 4
Between diagonals (i.e. interaction) sum of squares = 36

As a result of carrying out these three one-way analyses of variance we can extract the sum of squares due to each independent variable and their interaction. These are given in Table 15.9.

Table 15.9   Two-way analysis of variance table

Source of variation	Sum of squares	Degrees of freedom	Mean square
Between temperatures	4	1	4.0
Between feedrates	0	1	0.0
Interaction	36	1	36.0
Total	40	3	—

Comparing the sums of squares in Table 15.9 with those in Table 15.8 we can see that the 'within temperatures sum of squares' (36) cannot be classed as unassignable variation (due to sampling or testing error perhaps) but must be credited to the other independent variable (feedrate) and the interaction. Similarly the 'within feedrates sum of squares' (40) must be attributed to the effect of temperature and to the temperature × feedrate interaction. In fact *all* of the variation in yield must be credited to the three sources listed in Table 15.9 with *no residual* variation remaining. It is impossible therefore to test the mean squares in Table 15.9 by means of an *F*-test.

We had the same problem in Chapter 12, of course, when we attempted to analyse this data using the *t*-test. On that occasion we resorted to carrying out a second replicate of the 2^2 factorial experiment and the results of both replicates are given in Table 15.10.

Table 15.10   Yield from 2×2^2 factorial experiment (from Table 12.8)

Feedrate	Temperature (°C)	
	Low (60)	*High* (70)
Low (40)	77 75	73 71
High (50)	67 73	80 76

Now that we have more than one observation in each cell of Table 15.10 we could estimate the residual standard deviation to use in a *t*-test. In Chapter 12 we did just that. Alternatively we can use analysis of variance to subdivide the total variation into two components, one which can be assigned to the independent variables and one which we will call residual. This is done by using one-way analysis of variance with the eight yield values split into four groups such that all members of a group were produced under identical conditions. Each group corresponds to one of the four cells in Table 15.10. We will calculate the sums of squares by means of the formulae given earlier in this chapter.

Total sum of squares
= (overall SD)^2 (total number of observations − 1)
= $(3.9641)^2 (8 − 1)$
= 110.0

Within groups sum of squares
= Σ (group SD)^2 (number of observations in group − 1)
= $(1.4142)^2 (2 − 1) + (1.4142)^2 (2 − 1)$
 $+ (4.2426)^2 (2 − 1) + (2.8284)^2 (2 − 1)$
= 30.0

Between groups sum of squares
= (SD of group means)^2 (number of groups − 1)
 × (number of observations in each group)
= $(3.6515)^2 (4 − 1) (2)$
= 80.0

Table 15.11   One-way analysis of variance on 2×2^2 factorial experiment

Source of variation	Sum of squares	Degrees of freedom	Mean square
Between groups	80.0	3	26.67
Within groups	30.0	4	7.5
Total	110.0	7	—

Carrying out an *F*-test on the mean squares in Table 15.11 would show that the between groups mean square was not significantly greater than the within group mean square. The within groups mean square is a measure of the variability in yield that we would expect to find between batches produced with the same values of temperature and feedrate. The between groups mean square is also subject to this residual variation but may also be inflated by the effects of temperature and feedrate changes. Are we unable to conclude, then, that the temperature and feedrate effects exist? Despite the inconclusive *F*-test it would be unwise to abandon the analysis at this point for we can test the two main effects and the interaction separately. To do so we must break down the between groups sum of squares into three components (Table 15.12).

Table 15.12   Two-way analysis of variance on 2×2^2 factorial experiment

Source of variation	Sum of squares	Degrees of freedom	Mean square
Between temperatures	8.0	1	8.0
Between feedrates	0.0	1	0.0
Interaction	72.0	1	72.0
Between groups	80.0	3	—
Within groups (residual)	30.0	4	7.5
Total	110.0	7	—

The sums of squares above the broken line in Table 15.12 are calculated by considering the variation in the row means, column means, and the diagonal means of Table 15.10. *F*-tests can now be carried out to compare the mean squares above the line with the residual mean square and such tests would reveal that only the interaction between temperature and feedrate was significant. This is precisely the same conclusion that we reached in Chapter 12 when using the *t*-test.

15.6   Summary

In this chapter we have travelled a familiar route riding in a new vehicle, analysis of variance. It is hoped that you have recognized many landmarks first encountered in earlier chapters and that you have viewed the whole terrain from a new perspective. Perhaps any concepts which appeared disjoint when first encountered will now have assumed their rightful place in the whole ethos of design and analysis.

Problems

(1) The production manager of Indochem knows that the quality of his product is related to the concentration of a particular impurity in a liquid feedstock. Furthermore he suspects that this impurity gradually settles during storage of the feedstock. To investigate this possibility he asks an R & D chemist to take three samples of feedstock from a storage tank that has been undisturbed for several days. The samples are taken at depths of 0.5 m, 2.5 m and 4.5 m, then five determinations of impurity content are made on each sample.

Sample	Top	Middle	Bottom
Determinations of impurity	3.2 3.3 3.1 3.3 3.1	3.3 3.4 3.3 3.5 3.0	3.6 3.8 3.7 3.4 4.0

(a) Calculate the total sum of squares.
(b) Calculate the within samples sum of squares.
(c) Calculate the between samples sum of squares.
(d) Draw up a one-way analysis of variance table.
(e) Carry out an *F*-test to compare the between sample variation with the within sample variation.
(f) Calculate an estimate of the testing standard deviation.
(g) What conclusions can you draw concerning the settlement of the impurity within the tank?
(h) What other method of analysis would be appropriate in this situation?

(2) We could carry out a simple regression analysis on the data in Problem 1 using depth (x) as the independent variable and impurity (y) as the response. Feeding the 15 pairs of numbers into a regression analysis program results in the following equation being printed out:

$$y = 3.0875 + 0.1250x \qquad 57.87\% \text{ fit}$$

(a) Calculate the total sum of squares.
(b) Calculate the residual sum of squares.
(c) Calculate the regression sum of squares.
(d) Draw up an analysis of variance table.
(e) Carry out an *F*-test to check the statistical significance of the relationship between depth and impurity.

(3) We could carry out a multiple regression analysis on the data in Problem 1 using depth (x) and (depth)^2 as independent variables. Feeding the 15 pairs of numbers into a regression package, and asking for scaling of x into $X =$

$(x - 2.5)/2.0$ before the generation of X^2, we get the following equations printed out:

$$y = 3.4000 + 0.2500X \qquad\qquad 57.87\% \text{ fit}$$
$$y = 3.3000 + 0.2500X + 0.1500X^2 \qquad 64.815\% \text{ fit}$$

The analysis of variance table for the first equation will be identical with the one produced in problem 2(d). Scaling the independent variable has not changed either the standard deviation of y or the correlation between x and y, so the sums of squares are unchanged.

(a) Complete the analysis of variance table below:

Source of variation	Sum of squares	Degrees of freedom	Mean square
Due to regression on X			—
Due to introduction of X^2			
Due to regression on X and X^2			—
Residual			
Total			—

(b) Carry out an F-test on the mean squares in the above table.

(c) Compare the analysis of variance table from Problem 1 with the one produced in this problem and list any similarities.

———Appendix A———
The sigma (Σ) notation

Throughout this subject the calculation of certain statistics involves adding sets of numbers. The simplest case is where we add a set of observations. For example the first six batches of digozo blue gave the number of overloads as 0, 4, 2, 2, 1, 4 giving a total of 13. We can either refer to this total as 'the sum of the observations' or we can use a statistical shorthand which is far more concise. Thus Σx is the shorthand for 'the sum of the observations' where x is the symbol for an observation and Σ is the symbol for *add*. Therefore Σx is an instruction telling us to 'add the observations'.

$$\Sigma x = 0 + 4 + 2 + 2 + 1 + 4 = 13$$

We can now use this shorthand to represent other statistics. For example 'the sum of the squared observations' is denoted by Σx^2. Using the data given above:

$$\Sigma x^2 = 0^2 + 4^2 + 2^2 + 2^2 + 1^2 + 4^2 + = 41$$

For another example let us calculate $\Sigma (x-2)^2$. This gives:

$$\Sigma (x-2)^2 = (0-2)^2 + (4-2)^2 + (2-2)^2$$
$$+ (2-2)^2 + (1-2)^2 + (4-2)^2 = 13$$

The gain in simplicity and unambiguity can clearly be seen if we compare the mathematical expression with the written expression which is 'the sum of squares of the observations after two has been subtracted from each observation'.

———Appendix B———
Notation and formulae

Notation

n	sample size, number of observations or number of points
$\bar{x}$	sample mean
s	sample standard deviation
s^2	sample variance
p	sample proportion
μ	population mean
σ	population standard deviation
σ^2	population variance
π	population proportion (or probability)
z	standardized value
t	critical value from the t-table
c	a change (or difference) which we wish to detect
Σ	'the sum of' (see Appendix A)
r	$\begin{cases} \text{number of occurrences (in Part One)} \\ \text{correlation coefficient (in Part Two)} \end{cases}$
r_{xy}	correlation coefficient, between x and y
a b	$\left.\begin{array}{c} \text{intercept} \\ \text{slope} \end{array}\right\}$ of least squares regression line, $y = a + bx$
RSD	Residual standard deviation
k	number of independent variables in a regression equation

Formulae

Sample mean: $\bar{x} = \Sigma x/n$

Sample standard deviation:

$$s = \sqrt{[\Sigma(x-\bar{x})^2/(n-1)]} = \sqrt{[(\Sigma x^2 - n\bar{x}^2)/(n-1)]}$$

Poisson distribution: $\mu^r e^{-\mu}/r!$

Binomial distribution:

$$\frac{n!}{r!(n-r)!}\,(\pi)^r\,(1-\pi)^{n-r}$$

Normal distribution: $z = (x-\mu)/\sigma$

Confidence interval for μ: $\bar{x} \pm ts/\sqrt{n}$

Confidence interval for π: $p \pm t\sqrt{\left[\dfrac{p(1-p)}{n}\right]}$

Confidence interval for the difference between two population means:

$$|\bar{x}_1 - \bar{x}_2| \pm ts\sqrt{\left(\frac{1}{n_1} + \frac{1}{n_2}\right)}$$

Sample size needed to detect a change (c) in the population mean (μ): $n = (2ts/c)^2$

Sample size needed to estimate the population mean within $\pm c$: $n = (ts/c)^2$

Sample sizes needed to detect a difference (c) between two population means $(\mu_1$ and $\mu_2)$:

$$n_1 = n_2 = 2\left(\frac{2ts}{c}\right)^2$$

Sample sizes needed to estimate the difference between two population means within $\pm c$:

$$n_1 = n_2 = 2\left(\frac{ts}{c}\right)^2$$

Sample size needed to detect a change (c) in a population proportion: $n = (t/c)^2$

Sample size needed to estimate a population proportion within $\pm c$: $n = (t/2c)^2$

Localized standard deviation (for use in cusum test):

$$\sqrt{\left[\frac{\Sigma(x_c - x_{c+1})^2}{2(n-1)}\right]}$$

Covariance of x and y:

$$\Sigma(x - \bar{x})(y - \bar{y})/(n-1)$$
$$\text{or} \quad (\Sigma xy - -n\bar{x}\bar{y})/(n-1)$$

Correlation coefficient of x and y:

$$\frac{\text{covariance of } x \text{ and } y}{(\text{SD of } x)(\text{SD of } y)}$$

Least squares regression line $y = a + bx$

$$\text{Slope } b = \frac{\text{covariance of } x \text{ and } y}{\text{variance of } x}$$

$$\text{Intercept } a = \bar{y} - b\bar{x}$$

Significance test	Test statistic
One-sample t-test	$\dfrac{\lvert \bar{x} - \mu \rvert}{s/\sqrt{n}}$
t-test for two independent samples	$\dfrac{\lvert \bar{x}_1 - \bar{x}_2 \rvert}{s\sqrt{\left(\dfrac{1}{n_1} + \dfrac{1}{n_2}\right)}}$
	$s = \sqrt{\{[(n_1 - 1)\,s_1{}^2 + (n_2 - 1)\,s_2{}^2]/ [(n_1 - 1) + (n_2 - 1)]\}}$
t-test for two matched samples (paired comparison test)	$\dfrac{\lvert \bar{d} - \mu_d \rvert}{s_d/\sqrt{n}}$ ($\bar{d}$ and s_d are calculated from the *differences*)
F-test	Larger variance/smaller variance
One-sample proportion test	$\dfrac{\lvert p - \pi \rvert}{\sqrt{\left[\dfrac{\pi(1 - \pi)}{n}\right]}}$
Chi-squared test	$\sum \dfrac{(O - E)^2}{E}$
Dixon's test $(2 < n < 8)$	$\dfrac{x_2 - x_1}{x_n - x_1}$ or $\dfrac{x_n - x_{n-1}}{x_n - x_1}$
Dixon's test $(7 < n < 13)$	$\dfrac{x_2 - x_1}{x_{n-1} - x_1}$ or $\dfrac{x_n - x_{n-1}}{x_n - x_2}$
Dixon's test $(12 < n)$	$\dfrac{x_3 - x_1}{x_{n-2} - x_1}$ or $\dfrac{x_n - x_{n-2}}{x_n - x_3}$
Cusum test	$\dfrac{\text{maximum cusum}}{\text{localized standard deviation}}$
Regression t-test	$\sqrt{\{[(\text{new \% fit} - \text{old \% fit})\ (n - k - 1)]/[(100 - \text{new \% fit})]\}}$
Effect estimate from p replicates of a 2^n factorial experiment	$\dfrac{\text{effect estimate}}{\text{RSD}/\sqrt{(p2^{n-2})}}$

Residual $=$ actual value $-$ predicted value
Residual sum of squares:

$$\Sigma\,(\text{residual})^2$$

or
$$(1 - r^2)(n - 1)(\text{SD of } y)^2$$
with only one independent variable

Residual standard deviation (RSD):
$$\sqrt{[(\text{residual sum of squares})/(\text{residual degrees of freedom})]}$$

Residual degrees of freedom $= n - k - 1$

Confidence interval for true intercept (α) is:

$$a \pm t_{n-2} (\text{RSD}) \sqrt{\left[\frac{1}{n} + \frac{\bar{x}^2}{(n-1) \, \text{Var}(x)} \right]}$$

Confidence interval for true slope (β) is:

$$b \pm t_{n-2} (\text{RSD}) \sqrt{\left[\frac{1}{(n-1) \, \text{Var}(x)} \right]}$$

Confidence interval for the true value of y corresponding to a particular value of x (say X) is:

$$a + bX \pm t_{n-2} (\text{RSD}) \sqrt{\left[\frac{1}{n} + \frac{(X-\bar{x})^2}{(n-1) \, \text{Var}(x)} \right]}$$

Confidence interval for an observed value of y corresponding to a particular value of x (say X) is:

$$a + bX \pm t_{n-2} (\text{RSD}) \sqrt{\left[1 + \frac{1}{n} + \frac{(X-\bar{x})^2}{(n-1) \, \text{Var}(x)} \right]}$$

Percentage fit $= 100 \, r_{xy}^2$ (for $y = a + bx$)

Percentage fit $= 100(r_{xy}^2 + r_{zy}^2 - 2r_{xy}r_{zy}r_{xz})/(1 - r_{xz}^2)$
    (for $y = a + bx + cz$)

Percentage fit $= 100$ (regression sum of squares)/(total sum of squares)

──Appendix C──
Sampling distributions

It is suggested throughout this book that you can carry out significance tests or calculate confidence limits without being able to derive the formulae which are used in these activities. If you have already worked through the later chapters you will probably agree with this assertion. Nonetheless you may feel very uneasy about making use of formulae when you have little understanding of how they are derived and you might therefore wish to explore the theoretical basis of the methods advocated in this book.

Unfortunately a deeper understanding of statistical inference must *necessarily* be based on a knowledge of 'sampling distributions'. These are rather special forms of probability distributions and any discussion of them has been carefully avoided throughout this book. The reason why no mention has been made of sampling distributions is because they demand that you view the sampling process from a *very different* standpoint to that normally adopted by the scientist or technologist.

The scientist takes *one* sample of n items from a population and uses the information in the sample to infer certain characteristics of the population. *The whole purpose of the scientist's investigation is to learn more about the population.*

The mathematical statistician, on the other hand, starts with a population about which he knows everything and from this he takes *many* samples, each containing n items. *The purpose of this investigation is to learn more about how the mean (or some other statistic) varies from sample to sample.*

The contrast between the two approaches cannot be emphasized too strongly. The scientist/technologist takes *one* sample whereas the mathematical statistician takes *many* (perhaps even an infinite number). The scientist/technologist is operating in the *real world* whereas the mathematical statistician is operating in the *abstract world* of mathematics where sampling is effortless. To bridge the gap between these two worlds we will consider a hypothetical example in which the peculiar approach of the mathematical statistician is applied to a concrete situation. Please note that what follows is *not* a recommended strategy for the practical scientist.

We will start with a population which consists of a very large consignment of bottle tops. Imagine that the diameter of each bottle top is measured, then the measurement is written on a red ticket and the ticket placed in a large drum. Clearly we can regard the numbers on the red tickets as constituting a second population and we will refer to this as the *parent population*. Furthermore we will refer to the probability distribution of the numbers on the red tickets as the *parent distribution*. The situation is illustrated in Fig. C.1.

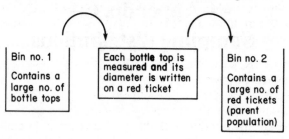

Figure C.1

From the parent population we now take a random sample of *n* red tickets, calculate the mean diameter for the sample ($\bar{x}$), write this sample mean on a blue ticket, place the blue ticket in a third bin and finally return the sample of red tickets to bin number 2 (Fig. C.2). This sampling procedure is repeated. As we are sampling with replacement we could continue indefinitely.

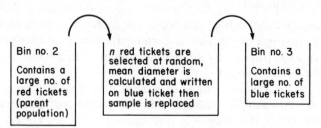

Figure C.2

The distribution of the numbers on the blue tickets is known as the *sampling distribution of the mean*. Clearly this sampling distribution tells us how the sample mean varies from sample to sample. The sample means in bin number 3 will have an average value which is known as the 'mean of the sampling distribution'. The sample means will be scattered around this average value and we can speak of the standard deviation of the sample means. It is the accepted convention to refer to this standard deviation as the '*standard error* of sample means'.

Many interesting questions can be asked concerning the relationship between the sampling distribution and the parent distribution. For example:

(a) How is the *mean* of the sampling distribution related to the mean of the parent distribution?

(b) How is the *standard error* of the sampling distribution related to the standard deviation of the parent distribution?
(c) How is the *shape* of the sampling distribution related to the shape of the parent distribution?

The mathematical statistician has given us answers to these questions. (His answers are based upon mathematics, not upon bottle tops or coloured tickets.) He tells us that, whatever the parent distribution:

(a) the mean of the 'sampling distribution of the mean' is equal to the mean of the parent distribution;
(b) the standard error of the 'sampling distribution of the mean' is equal to the standard deviation of the parent distribution (σ) divided by the square root of the sample size, i.e. $\sigma/\sqrt{n}$.

The standard error ($\sigma/\sqrt{n}$) will always be less than the standard deviation (σ) and will be very much less if n is large. This confirms what common sense would suggest, that sample means are not as widely scattered as individual observations, i.e. the mean diameters written on the blue tickets will have a smaller spread than the individual diameters written on the red tickets.

Concerning the *shape* of the sampling distribution of the mean it follows that it will be narrower than the parent distribution. The mathematical statistician further informs us that:

(a) the sampling distribution of the mean will be more like a normal distribution than is the parent distribution;
(b) the sampling distribution of the mean is closer to a normal distribution if n is large.

The relationship between the parent distribution and the sampling distribution is illustrated in Fig. C.3.

If the parent distribution is actually normal then the sampling distribution of the mean will also be normal. If the parent population does *not* have a normal distribution we can nonetheless be confident that the sampling distribution of the mean will be approximately normal provided n is 'large'. In a nutshell, if individual observations come from a population with mean μ and standard deviation σ then sample means can be considered to come from a population with mean μ and standard deviation (or standard error) $\sigma/\sqrt{n}$. From our knowledge of the normal distribution we can say that *95% of sample means will lie in the range $\mu \pm 1.96\sigma/\sqrt{n}$.*

From the point of view of the scientist who intends to take *one* sample we can say that there is a 95% chance that his sample mean will lie in the interval $\mu \pm 1.96\sigma/\sqrt{n}$. Much more important, we can look at the problem from the point of view of the scientist who has already taken *one* sample of n observations, and say that there is *a 95% chance that the population mean (μ) lies in the interval $\bar{x} \pm 1.96\sigma/\sqrt{n}$.* In Chapter 4 we referred to such intervals as 95% confidence intervals but we used a slightly different formula. This

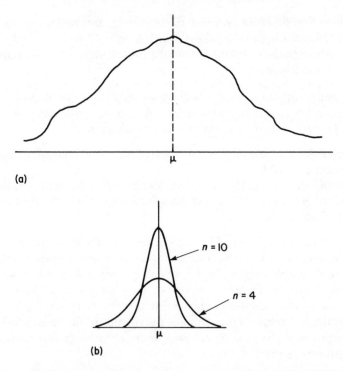

(a)

(b)

Figure C.3   (a) Parent distribution (mean μ, standard deviation σ) (b) Sampling distribution of the mean (mean μ, standard error $\sigma/\sqrt{n}$)

formulae ($\bar{x} \pm ts/\sqrt{n}$) is more useful because we would not normally know the value of the population standard deviation (σ) and would need to use the sample standard deviation (s). The extra uncertainty of using s instead of σ requires that the 1.96 be replaced by a critical value from the t-table. This t value will always exceed 1.96.

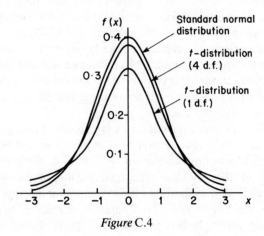

Figure C.4

The critical values in the t-table are actually obtained by finding areas under the *t-distribution* which is the sampling distribution of the statistic t [where $t = (\bar{x} - \mu)/(s/\sqrt{n})$]. Just as the sample mean ($\bar{x}$) varies from sample to sample so does t.

Mathematical statisticians have studied the sampling distribution of t in order to obtain the t-table. The t-distribution is rather similar to the standard normal distribution especially if n is large. This point is illustrated in Fig. C.4.

————————Appendix D————————
Copy of computer print-out from a multiple regression program

Data input

Number of cases	10	
Number of variables	6	
Variable names	Imp., Wt., Age, Feed, Temp., Agit.	

Imp.	Wt.	Age	Feed	Temp.	Agit.
4	3	1	3	1	3
3	4	2	5	2	3
4	6	3	7	3	3
6	3	4	5	3	2
7	1	5	2	2	2
2	5	6	6	1	2
6	1	7	2	2	2
10	0	8	1	3	4
5	2	9	3	1	4
3	5	10	6	2	4

Data analysis

Variable	Low	High	Mean	SD
Imp.	2	10	5.00	2.3570
Wt.	0	6	3.00	2.0000
Age	1	10	5.50	3.0277
Feed	1	7	4.00	2.0548
Temp.	1	3	2.00	0.81650
Agit.	2	4	2.90	0.87560

Correlation matrix

	Imp.	Wt.	Age	Feed	Temp.	Agit.
Imp.	1.00	−0.849	0.234	−0.780	0.520	0.108
Wt.	−0.849	1.00	−0.257	0.973	−0.068	0.000
Age	0.234	−0.257	1.00	−0.196	−0.045	0.440
Feed	−0.780	0.973	−0.196	1.00	0.066	−0.062
Temp.	0.520	−0.068	−0.045	0.066	1.00	0.00
Agit.	0.108	0.000	0.440	−0.062	0.00	1.00

Regression analysis

Dependent variable   Imp.

First equation

Variable	Coeff.
Wt.	−1.00
Const.	8.00

Percentage fit = 72.0%

Analysis of variance:

Source	SS	d.f.	MS
Regression on Wt.	36.0	1	36.0
Residual	14.0	8	1.75
Total	50.0	9	−

Partial correlation with dependent variable

Wt.	0.000
Age	0.030
Feed	0.378
Temp.	0.875
Agit.	0.203

Second equation

Variable	Coeff.
Wt.	−0.9628
Temp.	1.3395
Const.	5.2093

Percentage fit = 93.4%

Analysis of variance:

Source	SS	d.f.	MS
Regression on Wt.	36.00	1	–
Inclusion of Temp.	10.70	1	10.70
Regression on Wt. and Temp.	46.70	2	–
Residual	3.30	7	0.47
Total	50.0	9	–

No other independent variables are statistically significant.

─Appendix E─
Partial correlation

The correlation coefficient was introduced in Chapter 10 and is used extensively throughout Part Two of this book. It is a simple and very useful measure of the strength (and direction) of relationship between *two* variables. It is equally useful whether we are discussing the relationship between an independent variable and a dependent variable, or the relationship between two independent variables. It is not, however, quite so useful when we want to consider the relationships between *three* variables.

Suppose we have measured three variables; a dependent variable (y) and two independent variables $(x$ and $z)$. We can calculate three simple correlation coefficients, r_{xy}, r_{zy} and r_{xz}. If r_{xy} is greater than r_{zy} (ignoring any negative signs) then we would choose x as the first independent variable and fit the equation $y = a + bx$. If the magnitude of r_{xy} were greater than the critical values from Table H it would be reasonable to conclude that the variation in the independent variable, x, was reponsible for part of the variation in the dependent variable, y. Having made this decision we can now turn our attention to the second independent variable (z) and its relationship with the dependent variable. In doing so we must not ignore the possibility that *y may appear to depend on z simply because both y and z are dependent on x*; and with this possibility in mind it is reasonable to ask 'what would be the correlation between z and y if x did not vary?'

To answer this question we calculate what is known as the 'partial correlation between z and y with x constant' using the formula:

$$\frac{r_{yz} - r_{xz}r_{xy}}{\sqrt{[(1 - r_{xz}^2)(1 - r_{xy}^2)]}}$$

To illustrate the use of this formula we will use the data from Chapter 10 and concentrate on the three variables, weight of special ingredient (x), feedrate (z) and impurity (y). The relevant correlation coefficients taken from Appendix D are:

$$r_{xy} = -0.849 \qquad r_{zy} = -0.780 \qquad r_{xz} = 0.973$$

Calculation of the 'partial correlation between z and y with x fixed' proceeds as follows:

$$\frac{r_{yz} - r_{xz}r_{xy}}{\sqrt{[(1 - r_{xz}^2)(1 - r_{xy}^2)]}} = \frac{-0.780 - (0.973)(-0.849)}{\sqrt{[1 - 0.973^2)(1 - 0.849^2)]}}$$

$$= \frac{-0.780 + 0.826}{\sqrt{[(0.053)(0.279)]}}$$

$$= \frac{0.046}{0.122}$$

$$= 0.377$$

This result is in agreement with the partial correlation coefficient in the computer print-out of Appendix D. Note that the partial correlation between z and y (with x fixed) is *positive* whereas the simple correlation between z and y is *negative*. This is telling us that, *when the weight of special ingredient (x) is taken into account*, an increase in feedrate (z) can be expected to give an increase in impurity (y) and *not* a decrease in impurity as indicated by the simple correlation ($r_{zy} = -0.780$).

Though a partial correlation coefficient is much more complex than a simple correlation coefficient it is nonetheless just as easy to test the statistical significance of the former as the latter. We use the modulus of the partial correlation as the test statistic and we obtain the critical values from Table H, *but we must reduce the sample size by one*. Thus to test the partial correlation coefficient just calculated we would use a sample size of nine and the critical values would be 0.666 at the 5% level and 0.798 at the 1% level of significance. As the test statistic is equal to 0.377 we cannot reject the null hypothesis and we are unable to conclude that the percentage impurity (y) of a batch is dependent upon the feedrate (z). This conclusion is in agreement with that drawn in Chapter 11 after fitting the multiple regression equation $y = a + bx + cz$.

Clearly it is meaningless to speak of partial correlation when referring to a situation in which there are *only two* variables. We need a third, 'fixed' variable if we are to calculate a partial correlation coefficient. We are not, however, restricted to this level for the concept is useful when we have four or more variables. Suppose that we add a third independent variable, w, to the two already considered. After fitting the equation $y = a + bx + cz$ we would be interested in 'the partial correlation between y and w *with x and z fixed*'. This would be known as a second order partial correlation coefficient to distinguish it from the first order partial correlation in which we have only one fixed variable. When testing the significance of a second order coefficient we would again use critical values from Table H but we would reduce the sample size by *two*. When testing partial correlation coefficients of higher order we would reduce the sample size by subtracting the number of fixed variables.

A *part correlation coefficient* is rather similar to a partial correlation coefficient. With part correlation, however, the effect of the fixed variable is removed from *only one* of the other two variables. Thus we can calculate 'the part correlation of *y* and *z* with the effect of *x* removed from *z*' using:

$$\frac{r_{yz} - r_{xz}\, r_{xy}}{\sqrt{(1 - r_{xz}{}^2)}} = 0.200$$

Alternatively we can calculate 'the part correlation of *y* and *z* with the effect of *x* removed from *y*' using:

$$\frac{r_{yz} - r_{xz}\, r_{xy}}{\sqrt{(1 - r_{xy}{}^2)}} = 0.087$$

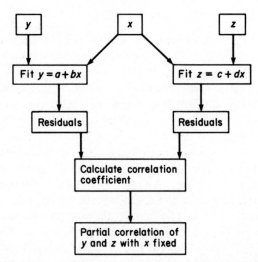

Figure E.1   Calculation of partial correlation coefficient

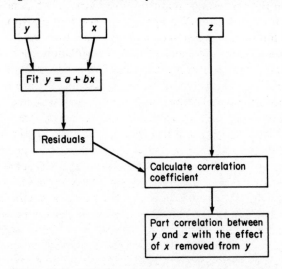

Figure E.2   Calculation of part correlation coefficient

Obviously there is a distinction between these two part correlation coefficients and the partial correlation coefficient calculated earlier. If you were restricted to using correlation rather than regression analysis the distinction would be very important. Hopefully you will have access to a multiple regression package and need only consider correlation for the light it sheds on the regression analysis. Perhaps the relation between the two will be made clear by the block diagrams, Figs E.1 and E.2, which illustrate an alternative method of calculation using residuals from regression equations. The important point to note is that 'removing the effect of x from y' can be achieved by fitting $y = a + bx$, then replacing the y values by the residuals from the equation.

--------------------------------**Appendix F**--------------------------------

Significance tests on effect estimates from a $p2^n$ factorial experiment

It was stated in Chapter 12, in a discussion of factorial experiments, that we can distinguish between real and chance effects by comparing:

$$\text{Test statistic} = \frac{\text{effect estimate}}{\text{RSD}/\sqrt{(p2^{n-2})}} \qquad \text{(F.1)}$$

with a critical value from the t-table. The foundation on which this test is based can be readily understood if we consider the way in which the effect estimates are calculated. For simplicity we will start with the 2^2 factorial design, which gives the four response values y_1, y_2, y_3 and y_4 in Table F.1.

Table F.1   Results of a 2^2 factorial experiment

	Temperature		
Feedrate	*Low*	*High*	*Mean*
Low	y_1	y_2	$\frac{1}{2}(y_1+y_2)$
High	y_3	y_4	$\frac{1}{2}(y_3+y_4)$
Mean	$\frac{1}{2}(y_1+y_3)$	$\frac{1}{2}(y_2+y_4)$	

To calculate an estimate of the feedrate main effect we subtract the mean response at the low level of feedrate from the mean response at the high level of feedrate:

$$\text{Feedrate main effect} = \frac{1}{2}(y_3+y_4) - \frac{1}{2}(y_1+y_2)$$
$$= \frac{1}{2}(y_3+y_4-y_1-y_2)$$

The other estimates can be expressed in a similar form:

$$\text{Temperature main effect} = \frac{1}{2}(y_2+y_4-y_1-y_3)$$
$$\text{Interaction effect} = \frac{1}{2}(y_1+y_4-y_2-y_3)$$

If we wish to test the statistical significance of the effect estimates we must first answer the question 'How would the effect estimates vary from experiment to experiment if the 2^2 factorial experiment were repeated many, many times?' It is clear that each response value would vary from experiment to experiment because each response value is subject to random error. This is made explicit in the model that was introduced in Chapter 12:

$$y_1 = \mu - \frac{T}{2} - \frac{F}{2} + \frac{I}{2} + e_1$$

$$y_2 = \mu + \frac{T}{2} - \frac{F}{2} - \frac{I}{2} + e_2$$

etc.

Within the model T, F and I are constants (although their values are unknown), whilst e_1, e_2, e_3 and e_4 are random errors. It is assumed that these random errors come from a normal distribution that has a mean equal to zero and a standard deviation equal to σ. (σ is also unknown and cannot be estimated from a simple factorial experiment, as we saw in Chapter 12.) It follows that:

$$SD(y_1) = \sigma \qquad SD(y_2) = \sigma \qquad SD(y_3) = \sigma \qquad SD(y_4) = \sigma$$

Note that, in speaking of the standard deviation of y_1, we are *not* referring to the variation amongst the response values within the experiment that was actually carried out. We are referring to the hypothetical variation, from experiment to experiment, of the response in the top left hand corner of Table F.1 (i.e. the yield of the batch made with low feedrate and low temperature). So, the variation of y_1 from experiment to experiment will have a standard deviation equal to σ and the same can be said of y_2, y_3 and y_4. We want to know how the effect estimates will vary from experiment to experiment, and each estimate is calculated from *four* response values. To help us with this step in the argument we will make use of the following well known results in mathematical statistics, in which 'Var' is an abbreviation for variance:

Theorem 1

$\text{Var}(x + z) = \text{Var}(x) + \text{Var}(z)$ if x and z are independent

Theorem 2

$\text{Var}(x - z) = \text{Var}(x) + \text{Var}(z)$ if x and z are independent

Theorem 3

$\text{Var}(kx) = k^2 \text{Var}(x)$ where k is a constant

Using these three theorems we can express the variance of each effect estimate in terms of σ^2. For example:

$$\text{Feedrate estimate (FE)} = \tfrac{1}{2}(y_3 + y_4 - y_1 - y_2)$$

$$\text{Var (FE)} = \tfrac{1}{4}\,\text{Var}\,(y_3 + y_4 - y_1 - y_2)$$

$$= \tfrac{1}{4}[\text{Var}\,(y_3) + \text{Var}\,(y_4) + \text{Var}\,(y_1) + \text{Var}\,(y_2)]$$

$$= \tfrac{1}{4}(\sigma^2 + \sigma^2 + \sigma^2 + \sigma^2)$$

$$= \sigma^2$$

Thus the standard deviation of the feedrate estimate is equal to the standard deviation of individual errors, i.e.

$$\text{SD (FE)} = \sigma$$

Similarly we could show that the standard deviation of the temperature estimate and of the interaction estimate were also equal to σ. To carry out a t-test on any of these effect estimates we calculate:

$$\text{Test statistic} = \frac{\text{effect estimate} - \text{true effect}}{s}$$

where s is an estimate of σ.

As our null hypothesis says that the true effect is equal to zero, and σ is estimated by a residual standard deviation (RSD), this becomes:

$$\text{Test statistic} = \frac{\text{effect estimate}}{\text{RSD}}$$

which is in keeping with equation (F.1) if we let $n = 2$ and $p = 1$.

Table F.2

Feedrate	Temperature		Mean
	Low	*High*	*Mean*
Low	y_1 y_5	y_2 y_6	$\tfrac{1}{4}(y_1 + y_5 + y_2 + y_6)$
High	y_3 y_7	y_4 y_8	$\tfrac{1}{4}(y_3 + y_7 + y_4 + y_8)$
Mean	$\tfrac{1}{4}(y_1 + y_5 + y_3 + y_7)$	$\tfrac{1}{4}(y_2 + y_6 + y_4 + y_8)$	

If we had carried out *two* replicates of a 2^2 factorial experiment to obtain the results in Table F.2 then the estimate of the feedrate main effect would be calculated from:

$$\text{Feedrate main effect} = \tfrac{1}{4}(y_3 + y_7 + y_4 + y_8) - \tfrac{1}{4}(y_1 + y_5 + y_2 + y_6)$$

$$= \tfrac{1}{4}(y_3 + y_7 + y_4 + y_8 - y_1 - y_5 - y_2 - y_6)$$

$$\text{Var (FE)} = \tfrac{1}{16}(\sigma^2 + \sigma^2 + \sigma^2 + \sigma^2 + \sigma^2 + \sigma^2 + \sigma^2 + \sigma^2)$$

$$= \tfrac{1}{2}\sigma^2$$

$$\text{SD (FE)} = \sigma/\sqrt{2}$$

Comparing this result with the corresponding result for a single replicate of the 2^2 factorial experiment we see that effect estimates from the larger experiment are more precise. For the 2×2^2 experiment it follows that the test statistic will be given by:

$$\frac{\text{Effect estimate}}{\text{RSD}/\sqrt{2}}$$

which is in agreement with equation (F.1) if we let $p = 2$ and $n = 2$.

Following similar arguments we could show that equation (F.1) gives the correct test statistic for any number of replicates of any 2^n factorial experiment.

Solutions to problems

Chapter 2

Solution to Problem 1

(a) Mean   4 (sum of ten values is 40)

      Median   3 (5th and 6th values are both 3)

(b) We know that the mean ($\bar{x}$) is 4.

The data values are:

x_i	4	2	7	3	0	3	1	13	4	3
$x_i - \bar{x}$	0	-2	3	-1	-4	-1	-3	9	0	-1
$(x_i - \bar{x})^2$	0	4	9	1	16	1	9	81	0	1

Therefore

$$\sum_{i=1}^{10} (x_i - \bar{x})^2 = 122$$

$$s^2 = \frac{\sum_{i=1}^{10} (x_i - \bar{x})^2}{n-1}$$

$$= \frac{122}{9}$$

$$\text{Variance} = 13.56$$

$$\text{Standard deviation} = \sqrt{13.56}$$

$$= 3.68$$

(c) Discrete.

Solution to Problem 2

(a) Mean $= \dfrac{\sum x_i}{n} = \dfrac{760}{8} = 95$

Median is 96.5 (4th and 5th values are 96 and 97 respectively).

To find the variance:

x_i	90	99	97	89	108	99	82	96
$x_i - \bar{x}$	-5	4	2	-6	13	4	-13	1
$(x_i - \bar{x})^2$	25	16	4	36	169	16	169	1

Therefore

$$\Sigma (x_i - \bar{x})^2 = 436$$

$$s^2 = \frac{\Sigma (x_i - \bar{x})^2}{n-1} = \frac{436}{7}$$

$$\text{Variance} = 62.29$$

$$\text{Standard deviation} = \sqrt{62.29}$$

$$= 7.89$$

$$\text{Coefficient of variation} = \frac{SD}{\text{mean}} \times 100$$

$$= \frac{7.89}{95} \times 100$$

$$= 8.30\%$$

(b) Continuous. (It seems that the experiment may well have been conducted by adding additional one pound weights to the skeins until they broke. Hence the data observed are in whole units. However, a skein which eventually broke under a load of 94 lb could well have broken under a load of 93.6 lb, or of 93.55 lb. The breaking strength is measured on a continuous scale.)

Solution to Problem 3

(a) Using the grouped frequency table:

With the 50 batches in the sample, the loss due to out-of-specification batches is:

No. of batches above specification = 3
Cost = 3 × £1000
Total cost = £3000

If the process mean had been reduced by 1.0 the sample mean would have been reduced from 245.49 to 244.49. The loss would then have been:

Cost of batches above specification = 1 × £1000 = £1000
Cost of batches below specification = 2 × £300 = £600
Total cost £1600

If the process mean had been reduced by 2.0 the sample mean would have been reduced from 245.49 to 243.49 and the loss would then have been:

Cost of batches below specification $= 4 \times £300$ $= £1200$
Total cost $= £1200$

Any further decrease will increase the number of batches below specification. It would appear that reducing the mean by 2.0 will minimize the loss.

We have, however, fallen into a major trap in decision-making.

We are interested in the *population* of batches and the sample has been taken to represent the population. Yet we have made the decision based on the sample completely disregarding the population. (The sample itself is of no interest since these batches have already been produced and their values cannot be altered.) What can we say about the population? One comment is that there is little information about how the population behaves at its extremities. It is, however, these extremes which are of major concern in this problem. Crosswell have two options to solve this problem:

(i) They obtain considerably more data. This could well be impossible.
(ii) They can fit a probability model as outlined in Chapter 3.

A realistic solution will therefore have to wait until after Chapter 3 has been undertaken. However, common sense will indicate that by reducing the process mean by 2.0, the lower specification limit is close to where the frequency of batches rapidly increases. It may well be prudent in the first instance to reduce the mean by less than 2.0.

(b) It is possible that this slight lack of symmetry in the histogram is due to sampling error. It is therefore quite likely that the distribution of the population is symmetrical.

(c) This is a situation in which it is possible to have more than one definition of the population. Two possible definitions are:

(i) All present and future batches. To make decisions about this population we have to assume that the first 50 batches are representative of the population.
(ii) All possible batches that theoretically could have been obtained during the period of production of the first 50 batches. The 50 batches are therefore a truly representative sample of this population but to use the information we have to assume the population will remain unchanged.

Chapter 3

Solution to Problem 1

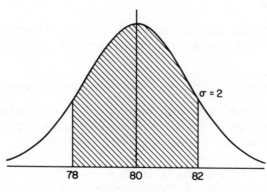

Figure S.1

(a) See Fig. S.1. Standardized value of $82 = Z_1 = (82-80)/2 = 1$.
Standardized value of $78 = Z_2 = (78-80)/2 = -1$.
Using Table A: 'area' to right of 1 is 0.1587.
Therefore also 'area' to left of -1 is 0.1587.
Therefore shaded 'area' is
$1.0 - (0.1587 + 0.1587) = 1.00 - 0.3174 = 0.6826$

Therefore 68.26% will have resistance of between 78 and 82 ohms.

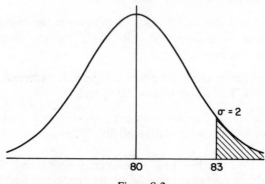

Figure S.2

(b) See Fig. S.2.
Standardized value of $83 = Z = (83-80)/2 = 1.5$.
Using Table A: 'area' to right of 1.5 is 0.0668.
Therefore 6.68% will have resistance of more than 83 ohms.

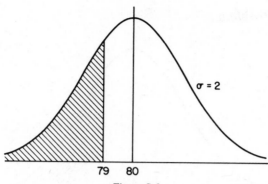

Figure S.3

(c) See Fig. S.3.
Standardized value of 79 = $Z = (79-80)/2 = -0.5$.
Using Table A: 'area' to right of 0.5 is 0.3085.
Therefore also 'area' to left of -0.5 is 0.3085.
Therefore P (resistor has resistance of less than 79 ohms) = 0.3085.

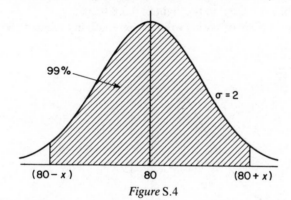

Figure S.4

(d) See Fig. S.4. For 99% between $(80-x)$ and $(80+x)$ we need 0.5% above $(80+x)$ and 0.5% below $(80-x)$. Using Table A (second part), we find that the standardization values of $(80+x)$ and $(80-x)$ must be $+2.58$ and -2.58; i.e.

$$2.58 = \frac{(80+x) - \text{mean}}{\text{standard deviation}}$$

$$= \frac{(80+x) - 80}{2}$$

$$= \frac{x}{2}$$

$$\therefore x = 5.16$$

Hence the limits are 74.84 ohms and 85.16 ohms.

Solution to Problem 2

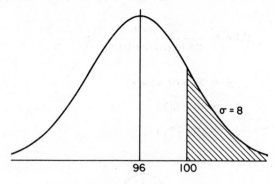

Figure S.5

(a) See Fig. S.5.
     Standardized value of $100 = Z = (100 - 96)/8 = 0.5$.
     Using Table A: 'area' to the right of 0.5 is 0.3085.
     Therefore 30.85% of skeins will have strength in excess of 100 lb.

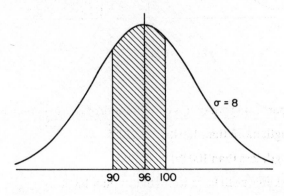

Figure S.6

(b) See Fig. S.6.
     Standardized value of $100 = Z_1 = (100 - 96)/8 = 0.5$.
     Standardized value of $90 = Z_2 = (90 - 96)/8 = -0.75$.
     Using Table A: 'area' to the right of 0.5 is 0.3085
                         'area' to the left of -0.75 is 0.2266.
     Therefore P (strength between 90 and 100 lb)

$$= 1 - 0.3085 - 0.2266$$

$$= 0.4649$$

(c) See Fig. S.7. Using Table A (in 'reverse' direction), the standardized value which has an 'area' of 0.2 to the right of itself is (approx.) 0.84; i.e.

$$0.84 = \frac{x - \text{mean}}{\text{standard deviation}} = \frac{x - 96}{8}$$

$$\therefore x = 96 + (0.84 \times 8)$$
$$= 96 + 6.72$$
$$= 102.72 \text{ lb}$$

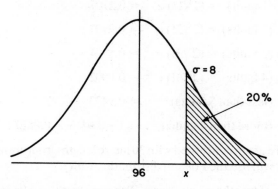

Figure S.7

(d) From (a) percentage of skeins with strength of 100 lb or more is 30.85%; i.e.

P (strength more than 100 lb) $= 0.3085$

$\therefore P$ (strength less than 100 lb) $= 0.6915$

$\therefore P$ (two skeins both have strength less than 100 lb)

$$= (0.6915)^2$$
$$= 0.4782$$

Solution to Problem 3

This is a situation in which the Poisson distribution might be applicable. By examination of the manufacturer's records we could establish the type of distribution that would best match the distribution of spotting faults in 15 metre lengths. Since we are unable to do this we will use a Poisson distribution with an appropriate value of μ.

(a) We will use a Poisson distribution with $\mu = 1$, since the fault rate is 1 fault per 15 metres when the process is working normally.

$$\text{Prob. of } r \text{ faults} = (\mu^r/r!)e^{-\mu}$$

$$\text{Prob. of } 0 \text{ faults} = (1^0/0!)e^{-1}$$

$$= 0.3679$$

(b) As we are now taking a 30 metre length we use a Poisson distribution with $\mu = 2$.

(i) Prob. (0 faults) $= (2^0/0!)\,e^{-2} = 0.1353$

(ii) Prob. (1 fault) $= (2^1/1!)\,e^{-2} = 0.2707$

Prob. (2 faults) $= (2^2/2!)\,e^{-2} = 0.2707$

Prob. (3 faults) $= (2^3/3!)\,e^{-2} = 0.1804$

Prob. (4 faults) $= (2^4/4!)\,e^{-2} = 0.0902$

Prob. (less than 5 faults) $= 0.9473$

Prob. (more than 4 faults) $= 1 - 0.9473 = 0.0527$

(c) (i) We are again concerned with 30 metre lengths from normal production so we will use the Poisson distribution with $\mu = 2$.

Prob. of less than 5 faults in a 30 metre length $= 0.9473$.

Prob. that a normal roll will be subjected to full inspection

$$= 0.9473\,(1 - 0.9473) + (1 - 0.9473)\,0.9473 + (1 - 0.9473)^2$$

$$= 0.1026$$

(ii) The number of faults in 30 metre lengths taken from a roll with a fault rate of one occurrence of spotting per 5 metres of paper will have a Poisson distribution with $\mu = 6$.

Prob. (0 faults) $= 0.0025$
Prob. (1 fault) $= 0.0149$
Prob. (2 faults) $= 0.0446$
Prob. (3 faults) $= 0.0892$
Prob. (4 faults) $= 0.1339$

$$\overline{0.2851}$$

Prob. of less than 5 faults in a 30 metre length taken from a roll with one fault per 5 metres $= 0.2851$

Probability of such a roll avoiding full inspection

$$= (0.2851)^2 = 0.0813$$

Solution to Problem 4

(a) The higher the mean weight setting the smaller will be the percentage of packets which have net weights less than 500 grams. Since the packet weights have a normal distribution the percentage of underweight packets cannot be reduced to zero however high we set the mean. Unfortunately we cannot comply with this regulation unless we weigh *every* packet individually and reject the underweight packets.

There is a movement away from wording laws in absolute terms such as '*no* package should have a weight which is less than the nominal weight', to wording laws in probabilistic terms.

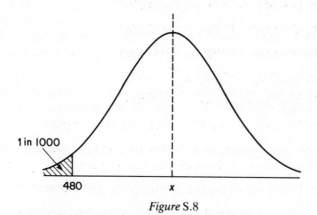

1 in 1000

480   x

Figure S.8

(b) See Fig. S.8. For a tail probability of 0.001 the standardized value is 3.10 (Table A).

$$\text{Hence} \quad -3.10 = \frac{480 - x}{10}$$

$$x = 480 + 31.0$$

$$= 511 \text{ grams}$$

The process mean should be set to 511 grams at least in order to ensure that there is less than 1 in 1000 chance of a packet containing less than 480 grams.

(c) See Fig. S.9. For a tail probability of 0.025 the standardized value is 1.96

$$\text{Hence} \quad -1.96 = \frac{500 - x}{10}$$

$$x = 500 + 19.6$$

$$= 519.6$$

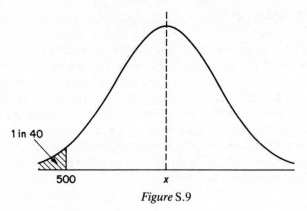

Figure S.9

The process mean should be set to 519.6 grams at least in order to ensure that there is less than 1 in 40 chance of a packet containing less than 500 grams.

(d) The mean weight should be set at a figure at least as great as the greater of the two values calculated in (b) and (c). A suitable value would be 520 grams.

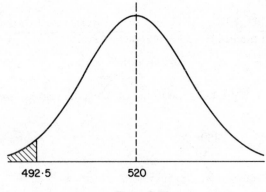

Figure S.10

(e) With the process mean set at 520 grams:

  (i) Rule 1 will be satisfied as the mean is greater than 500.
 (ii) See Fig. S.10.

$$\text{Standardized value} = \frac{492.5 - 520}{10}$$

$$= -2.75$$

Thus 99.7% of packages would have weights above the tolerance limit and Rule 2 is satisfied.

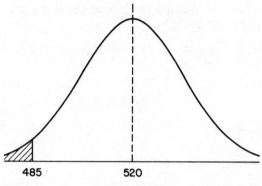

Figure S.11

(iii) Rule 3 cannot be satisfied because of the infinitely long tail of the normal distribution.

(f) See Fig. S.11.

$$\text{Standardized value} = \frac{485 - 520}{10}$$

$$= -3.5$$

Thus only 99.98% of packages would have weights above the absolute tolerance limit and Rule 3 would be violated.

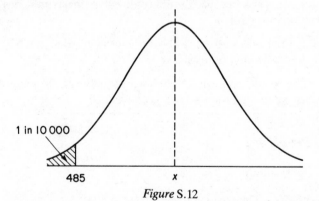

1 in 10 000

485   *x*

Figure S.12

(g) The mean must be increased to a value greater than 520 grams if Rule 3 is to be satisfied (see Fig. S.12).

$$-3.72 = \frac{485 - x}{10}$$

$$x = 485 + 37.2$$

$$= 522.2 \text{ grams}$$

Smith and Co. must increase the process mean weight to 522.2 grams in order to comply with Rule 3.

Thus the minimum setting for the process mean weight which will satisfy all three rules is 522.2 grams.

Chapter 4

Solution to Problem 1

$n = 6$ $\bar{x} = 6.27$ $s = 0.0645$

Null hypothesis: $\mu = 6.20$ True mean of consignment is 6.2.

Alternative hypothesis: $\mu > 6.20$

Test statistic $= \dfrac{|\bar{x} - \mu|}{s/\sqrt{n}}$

$= \dfrac{|6.27 - 6.20|}{0.0645/\sqrt{6}}$

$= 2.66$

Critical value – t-table with 5 ($n - 1$) degrees of freedom with a one-sided risk gives value of 2.02 and 3.36 at 5% and 1% significance levels respectively. We use a *one-sided* risk because the alternative hypothesis indicates a change from 6.20 in *one* direction.

Decision – Since the test statistic is above the critical value at a 5% level we accept the alternative hypothesis.

Conclusion – The results confirm, at a 5% significant level, the manufacturer's claim that the mean intrinsic viscosity is above 6.20.

Solution to Problem 2

(a) $n = 8$ $\bar{x} = 49.60$ $s = 0.968$

Null hypothesis – The true mean is 50.0 ($\mu = 50.0$).

Alternative hypothesis – The true mean is *not* 50.0 ($\mu \neq 50.0$).

Test statistic $= \dfrac{|\bar{x} - \mu|}{s/\sqrt{n}}$

$= \dfrac{|49.6 - 50.0|}{0.968/\sqrt{8}}$

$= 1.17$

Critical value – t-table with 7 degrees of freedom and a two-sided risk gives a value of 2.36 at a 5% significance level.

Decision – Since the test statistic is less than the critical value we cannot reject the null hypothesis.

Conclusion – There is no evidence that the method is biased at a 5% significance level.

(b) The 95% confidence interval is given by:

$$\bar{x} \pm \frac{ts}{\sqrt{n}} = 49.60 \pm \frac{2.36 \times 0.968}{\sqrt{8}}$$

$$= 49.60 \pm 0.81$$

$$= 48.79 \text{ to } 50.41$$

The maximum bias, at a 95% confidence level, is $(50 - 48.79)$ 1.21.

(c) If this is an initial evaluation it may well be carried out by a single highly skilled operator on the same day. The population for this situation is 'all possible values that could be obtained by the operator on that day'.

If, however, the method is shortly to become a standard procedure in the laboratory the population of interest is 'all possible values that could be obtained by all possible operators under all possible conditions'. The sample will have to be reasonably representative of this population.

Solution to Problem 3

(a) $n = 10$ $\bar{x} = 4.5$ $s = 1.144$

Null hypothesis – The true mean for the modified process is 5.4 ($\mu = 5.4$).

Alternative hypotheses: $\mu < 5.4$

Test statistic $= \dfrac{|\bar{x} - \mu|}{s/\sqrt{n}}$

$\qquad\quad = \dfrac{|4.5 - 5.4|}{1.144/\sqrt{10}}$

$= 2.49$

Critical value – t-table one-tailed, 9 degrees of freedom gives 1.83 and 2.82 at 5% and 1% significance levels.

Decision – We reject the null hypothesis and accept the alternative at the 5% significance level.

Conclusion – There is evidence, at the 5% significance level, that the mean impurity has decreased.

(b) t is 2.26 with 9 degrees of freedom and a two-sided risk.

95% confidence interval for the mean is given by:

$$\bar{x} \pm ts/\sqrt{n} = 4.5 \pm 2.26 \times 1.144/\sqrt{10}$$

$$= 4.5 \pm 0.82$$

$$= 3.68 \text{ to } 5.32$$

(c) Sample size $= \left(\dfrac{ts}{c}\right)^2$

$$= \left(\dfrac{2.26 \times 1.144}{0.15}\right)^2$$

$$= 297.08$$

Rounding to the next highest integer gives a sample size of 298.

Chapter 5

Solution to Problem 1

$n_A = 10 \qquad s_A = 2.951$

$n_B = 9 \qquad s_B = 1.323$

Null hypothesis – The population variances are equal ($\sigma_A^2 = \sigma_B^2$).

Alternative hypothesis – The population variances are unequal ($\sigma_A^2 \neq \sigma_B^2$).

$$\text{Test statistic} = \frac{\text{sample variance of A}}{\text{sample variance of B}}$$

$$= \frac{2.951^2}{1.323^2}$$

$$= 4.98$$

Critical value – With 9 degrees of freedom for the numerator and 8 for the denominator the F-table gives values of 4.36 and 7.34 at the 5% and 1% levels respectively for a two-sided test.

Decision – Since the test statistic is greater than the critical value at a 5% significance level we reject the null hypothesis and accept the alternative.

Conclusion – From this evidence it would appear that method B is more repeatable but it must be emphasized that the population consists of determinations from *one* operator and *one* sample of cement and therefore the conclusion is of limited applicability.

Solution to Problem 2

$n_A = 6$ $\bar{x}_A = 34.6$ $s_A = 6.046$

$n_w = 6$ $\bar{x}_w = 33.4$ $s_w = 5.476$

(a) Null hypothesis – The population variances are equal ($\sigma_A^2 = \sigma_w^2$).

   Alternative hypothesis – The population variances are unequal ($\sigma_A^2 \neq \sigma_B^2$).

$$\text{Test statistic} = \frac{\text{variance of A}}{\text{variance of W}}$$

$$= (6.046)^2/(5.476)^2$$

$$= 1.22$$

Critical value – F-table with a two-sided test, 5 and 5 degrees of freedom, gives 7.15 and 14.94 at the 5% and 1% levels respectively.

Decision – We cannot reject the null hypothesis at a 5% significance level.

(b) Since there is no evidence that the variances are unequal we can now combine them to give a better estimate of the population standard deviation.

$$s^2 = \frac{\text{d.f.}_A s_A^2 + \text{d.f.}_w s_w^2}{\text{d.f.}_A + \text{d.f.}_w}$$

$$= \frac{5 \times 6.046^2 + 5 \times 5.476^2}{5 + 5}$$

$$= 33.270$$

$$s = 5.768$$

Null hypothesis – The population means are equal ($\mu_A = \mu_w$).

Alternative hypothesis – The population mean for the oil containing additive is greater than the population mean for oil without additive ($\mu_A > \mu_w$).

$$\text{Test statistic} = \frac{(\bar{x}_A - \bar{x}_w)}{s\sqrt{\left(\dfrac{1}{n_A} + \dfrac{1}{n_w}\right)}}$$

$$\frac{(34.6 - 33.4)}{5.768\sqrt{\left(\dfrac{1}{6} + \dfrac{1}{6}\right)}}$$

$$= 0.36$$

Critical value – t-table, one-sided risk, with 10 degrees of freedom $(5 + 5)$ gives values of 1.81 and 2.76 at the 5% and 1% level respectively.

Decision – We cannot reject the null hypothesis at a 5% significance level.

Conclusion – There is no evidence to suggest that the additive decreases petrol consumption.

(c) t with a one-sided risk and 10 degrees of freedom has a value of 2.20.
    $s = 5.768, c = 1.0$.

$$n_1 = n_2 = 2 \left(\frac{2ts}{c} \right)^2$$

$$= 2 \left(\frac{2 \times 1.81 \times 5.768}{1.0} \right)^2$$

$$= 871.6$$

1744 cars ($n_1 = n_2 = 872$) are required to be 95% certain of detecting an improvement of 1.0 m.p.g.

(d) The company presumably would like the population to be 'All cars being used in the UK at the present time'.

Obtaining a random sample would necessitate that each car had an equal chance of being chosen in the sample. Clearly this would be impossible. However, by taking into account factors like engine size, car size and popularity, a reasonably representative sample can be chosen by selecting certain makes of cars.

Solution to Problem 3

(a) d is the increase in m.p.g.

Car	A	B	C	D	E	F
d	1.6	1.1	1.7	0.3	0.3	2.2

$\bar{x}_d = 1.2 \qquad s_d = 0.780$

Null hypothesis $\qquad \mu_d = 0$

Alternative hypothesis $\qquad \mu_d > 0$

Test statistic $= \dfrac{|\bar{x}_d - \mu_d|}{s_d/\sqrt{n}}$

$$= \frac{|1.2 - 0|}{0.780/\sqrt{6}}$$

$$= 3.77$$

Critical value – *t*-table, one-sided risk, with 5 degrees of freedom gives values of 2.02 and 3.36 at the 5% and 1% levels respectively.

Decision – We accept the alternative hypothesis at a 1% significance level.

Conclusion – There is substantial evidence of a decrease in petrol consumption.

(b)
$$n = \left(\frac{2ts}{c}\right)^2$$

t (one-sided, 5%, 5 d.f.) = 2.02, s = 0.780, c = 1.0

$$n = \left(\frac{2 \times 2.02 \times 0.78}{1.0}\right)^2$$

$$= 9.92$$

$$= 10$$

A sample size of ten cars is needed to be 95% certain of detecting an increase in 1.0 m.p.g. Since there are two trials per car, a total of 20 trials are needed. This is a considerable improvement on the previous design which required 1744 cars.

Chapter 6

Solution to Problem 1

(a) Null hypothesis – The colour-matcher cannot discriminate between the fabrics.

Alternative hypothesis – The colour-matcher can discriminate between the fabrics.

Test statistic = 7.

Critical value – The table of critical values for triangular test gives 8½ at a 5% level and 9½ at a 1% level.

Decision – We cannot reject the null hypothesis. It has not been proved that the colour-matcher can discriminate.

(b) In this situation it is possible that the colour-matcher could make the 'right' decision for the wrong reason. For example he may choose correctly the odd-one-out in the belief that it is lighter when it is in fact darker. A better design would be to present him with one sample from each fabric and ask him to choose the darker. The triangular design is only suitable when it is believed that there is no difference between fabrics.

Solution to Problem 2

(a) Null hypothesis $\pi = 0.10$

Alternative hypothesis $\pi > 0.10$

$$\text{Test statistic} = \frac{|p - \pi|}{\sqrt{[\pi(1 - \pi)/n]}}$$

$$(p = 31/200 = 0.155)$$

$$= \frac{|0.155 - 0.10|}{\sqrt{[0.1 \times 0.9/200]}}$$

$$= 2.59$$

Critical value – t-table with a one-sided risk and infinite degrees of freedom gives 1.64 at 5% and 2.33 at 1%.

Decision – Accept the alternative hypothesis at a 1% significance level.

Conclusion – There is evidence at the 1% significance level that the furnace has deteriorated in performance.

(b) 95% confidence interval for the true proportion is given by:

$$p \pm t \sqrt{[p(1 - p)/n]} = 0.155 \pm 1.96 \sqrt{[0.155 \times 0.845/200]}$$

$$= 0.155 \pm 0.050$$

$$= 0.105 \ (10.5\%) \text{ to } 0.205 \ (20.5\%)$$

(c) Every process is subject to random perturbations which result in each item being slightly different. In this process every insulator will be different and any slight change in conditions would have resulted in the random perturbations combining to give a different set of 200 insulators. We can therefore consider the day's actual production as a sample of 200 from all the insulators that *could* have been produced on that day.

(d) Observed values

Insulator	Inspector A	B	Total
Hazed	8	23	31
Not-hazed	72	97	169
Total	80	120	200

(e) Expected value (EV) $= \dfrac{\text{row total} \times \text{column total}}{\text{grand total}}$

For hazed insulators by Inspector A (EV) $= \dfrac{31 \times 80}{200} = 12.4$

All the other values can be obtained by subtraction from row and column totals to give:

	Inspector		
Insulator	A	B	Total
Hazed	12.4	18.6	31
Not-hazed	67.6	101.4	169
Total	80	120	200

Null hypothesis – No relationship between insulator classification and inspector ($\pi_A = \pi_B$).

Alternative hypothesis – Insulator classification related to inspector ($\pi_A \neq \pi_B$).

Test statistic $= \Sigma \dfrac{(O - E)^2}{E}$

$$= \dfrac{(8 - 12.4)^2}{12.4} + \dfrac{(23 - 18.6)^2}{18.6} + \dfrac{(72 - 67.6)^2}{67.6} + \dfrac{(97 - 101.4)^2}{101.4}$$

$$= 3.079$$

Critical value – Chi-squared table with 1 degree of freedom gives values of 3.841 and 6.635 at 5% and 1% significance levels respectively.

Degrees of freedom $=$ (rows $- 1$) $\times$ (column $- 1$)
$$= 1 \times 1 = 1$$

Decision – There is not enough evidence to reject the null hypothesis at a 5% significance level.

Conclusion – Although there is no significant difference between proportions at a 5% level there is a significant difference at a 10% level (critical value 2.69). There is a possibility that the inspectors give different proportions but more evidence will be needed to confirm this.

(f) Prob. (r hazings) $= \dfrac{n!}{r!(n-r)!} \pi^r (1 - \pi)^{n-r}$

$$(r = 1, n = 10, \pi = 0.1)$$

$$= \dfrac{10!}{1! \, 9!} \times 0.1 \times 0.9^9$$

$$= 10 \times 0.1 \times 0.387$$

$$= 0.387$$

Prob. (5 and
     above) $1.0 - (0.349 + 0.387 + 0.194 + 0.057 + 0.011)$

                     $= 0.002$

Prob. (2 or more
     hazings) $= 0.194 + 0.057 + 0.011 + 0.002$

                     $= 0.264$

(g) $r = 0$ $n = 10$ $\pi = 0.2$

   Prob. (0 hazings) $= \dfrac{10!}{0!\ 10!} \times 0.2^0 \times 0.8^{10}$ $(0! = 1, 0.2^0 = 1)$

                     $= 0.8^{10}$

                     $= 0.107$

   Prob. (4 hazings) $= 1.0 - (0.107 + 0.268 + 0.302 + 0.201 + 0.034)$

                     $= 0.088$

Prob. (less than 2
     hazings) $= 0.107 + 0.268$

                     $= 0.375$

(h) Using the results from part (g):

  (i) 0.264
  (ii) 0.375

Both these probabilities are high and there is a large chance of making a wrong decision. This is not too surprising since ten is a very small sample, when the only information from each sample is whether it is defective or not. A sample size of ten can, however, be adequate for a measured variable.

Chapter 7

Solution to Problem 1

(a) Rearranging the results in order of magnitude gives:

    27.3   30.7   31.7   32.3   32.4   32.5   32.7   34.1

With eight results, Dixon's test statistic uses the formulae:

$$\frac{x_2 - x_1}{x_{n-1} - x_1} \quad \text{and} \quad \frac{x_n - x_{n-1}}{x_n - x_2}$$

With the lowest value this gives:

$$\frac{30.7 - 27.3}{32.7 - 27.3} = \frac{3.4}{5.4} = 0.629$$

With the highest value this gives:

$$\frac{34.1 - 32.7}{34.1 - 30.7} = \frac{1.4}{3.4} = 0.411$$

The critical values for a sample of eight are 0.608 and 0.717 for 5% and 1% significance levels respectively. We can therefore conclude that the value of 27.3 is an outlier at the 5% significance level.

(b) The rejection (or not) of the outlier will depend upon the answers to two questions:

  (i) Are the assumptions underlying the outlier test valid in this situation? Dixon's test is dependent upon the distribution being normal. We have no information regarding this assumption but it is highly likely that the information will be available to the car manufacturer from previous tests.

  (ii) How are the results to be used? If the data usually follow a normal distribution and if a confidence interval is to be calculated it is preferable that the outlier be rejected. However, the outlier might be the most important result since it could be indicative of a major fault in the car's design.

Solution to Problem 2

(a) Combined SD $= \sqrt{\left(\dfrac{\Sigma \ \text{d.f.} s^2}{\Sigma \ \text{d.f.}}\right)}$

$$= \sqrt{\ [(2 \times 0.192^2 + 2 \times 0.032^2 + 2 \times 0.035^2 + 2 \times 0.050^2}$$
$$+ 2 \times 0.380^2 + 2 \times 0.069^2)/(2 + 2 + 2 + 2 + 2 + 2)]}$$

$$= 0.178$$

(b) 95% confidence interval is given by:

$$\bar{x} \pm ts/\sqrt{n}$$

$\bar{x} = 1.93$, t (12 degrees of freedom) $= 2.18$, $s = 0.178$, $n = 3$.

$$1.93 \pm 2.18 \times 0.178/\sqrt{3}$$

$$1.93 \pm 0.224$$

$$1.71 \text{ to } 2.15$$

(c)

| Laboratory | Test statistic | |
	Smallest value	Largest value
A	0.368	0.632
B	0.167	0.833
C	0.000	1.000
D	0.400	0.600
E	0.871	0.129
F	1.000	0.000

Critical value at a 5% significance level is 0.970.

Therefore the largest value for laboratory C and smallest value for laboratory F could be classed as outliers.

(d) The combined SD is unduly high compared with the individual SDs for laboratories B, C, D and F. This is due to the large influence of laboratory E on the calculation.

The 95% confidence interval for laboratory B is unduly wide, owing to the use of a rather high combined SD.

A visual examination of the results would lead us to believe that the laboratories A and E might have an outlier among their results. Nobody would expect an outlier in laboratories C and F.

In conclusion, none of the three analyses seems fairly to reflect the pattern of results.

(e) (i) There are probably too few determinations to draw a conclusion. However, previous experience indicates that the normal distribution is usually valid for analytical determinations.
    (ii) An examination of the standard deviations reveals great differences between them. An outlier test for standard deviations (Cochran's test) would indicate that the laboratories A and E are outliers as far as variability is concerned. The assumption of equal standard deviations has been violated.
   (iii) To obtain independent determinations it is necessary that the value of the second (or third) determination should not be related to the first (or second) determination in each laboratory. A repeat determination should therefore result from a repeat of full analytical procedure and not just from part of it. It is also necessary that an operator is unaware of the value of previous determinations on the same sample. Hopefully the assumption of independent observations has not been violated.

(iv) Clearly we cannot obtain a random sample but hopefully our sample is representative of the personnel and procedures that will be used in future determinations.

(v) A continuous variable has been used but the results are rounded to two decimal places. It is this rounding that has resulted in outliers being found in laboratories C and F.

Solution to Problem 3

(a) $\bar{x} = 31.5$, $s = 28.01$, $n = 4$, t with 3 degrees of freedom $= 3.18$.

$$\bar{x} \pm ts/\sqrt{n} = 31.5 \pm 3.18 \times 28.01/\sqrt{4}$$
$$= 31.5 \pm 44.5$$
$$= -13.0 \text{ to } 76.0$$

(b) $p = 0.5$, $t = 1.96$, $n = 4$.

$$p \pm t\sqrt{[p(1-p)/n]} = 0.5 \pm 1.96\sqrt{(0.5 \times 0.5/4)}$$
$$= 0.5 \pm 0.49$$
$$= 0.01 \text{ to } 0.99$$

(c) (i) If the population is 'the time period under review' the sample might be considered as representative of the population. However, only at certain times during this period was the subject in contact with radiation. It would therefore appear that the population is not homogeneous and to draw inferences from the sample in relation to a single population would be dubious.

(ii) After the first result a decision was made to remove the subject from contact with radiation. Therefore the second result cannot be independent of the first.

(iii) Even with only four results there is an indication that the results do not follow a normal distribution.

(iv) The number of 'successes' is only two which is below the minimum of five.

(d) The fact that the samples are not independent invalidates both the confidence interval of the mean and the confidence interval of the proportion. The confidence interval of the mean is also invalid due to non-normality whereas the confidence interval for the proportion does not fill a necessary condition for use of the formula in part (b). The nature of the population under consideration is also unclear.

Chapter 9

Solution to Problem 1

(a) The mean mileage is 124.

(b)

Purchase	Mileage	Deviation from mean	Cusum	Purchase	Mileage	Deviation from mean	Cusum
1	135	11	11	21	104	−20	74
2	123	−1	10	22	123	−1	73
3	134	10	20	23	115	−9	64
4	141	+17	37	24	103	−21	43
5	127	3	40	25	103	−21	22
6	126	2	42	26	122	−2	20
7	132	8	50	27	115	−9	11
8	120	−4	46	28	131	7	18
9	122	−2	44	29	120	−4	14
10	141	+17	61	30	111	−13	1
11	129	+5	66	31	119	−5	−4
12	121	−3	63	32	126	2	−2
13	148	+24	87	33	134	+10	8
14	137	+13	100	34	126	+2	10
15	132	+8	108	35	128	+4	14
16	136	+12	120	36	110	−14	0
17	120	−4	116	37	114	−10	−10
18	103	−21	95	38	122	−2	−12
19	116	−8	87	39	123	−1	−13
20	131	+7	94	40	137	+13	0

For graph see Fig. S.13.

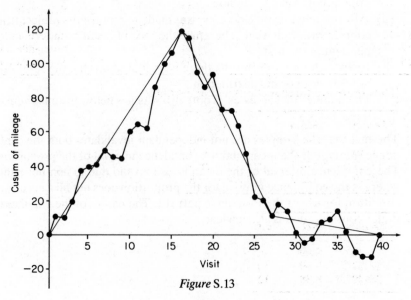

Figure S.13

(c) Null hypothesis – There was no change in the mean mileage during observations 1–40.

Alternative hypothesis – There was a change in mean mileage during this
period.

$$\text{Test statistic} = \frac{\text{maximum cusum}}{\text{combined SD}}$$

$$= \frac{120}{7.83}$$

$$= 15.32$$

Critical value – For a span of 40 observations, see table:

8.0 at the 5% significance level
9.3 at the 1% significance level.

Decision – reject the null hypothesis at the 1% significance level.

Conclusion – There has been a change in mean mileage about the time of
observation 16.

A further series of significance tests need to be carried out. These are
summarized below:

Section	Observation number for maximum cusum	Maximum cusum	Test statistic	Critical value 5%	Critical value 1%	Decision
1–40	16	120	15.32	8.0	9.3	Reject n.h.
1–16	12	27	3.44	4.7	6.0	Not significant
17–40	27	54	6.90	5.9	7.2	Reject n.h.
17–27	23	13.6	1.74	3.8	5.1	Not significant
28–40	39	12.2	1.56	4.2	5.5	Not significant

Thus there are three sections within the data:

Section	Mean	m.p.g.
1–16	131.5	131.5/4 = 32.9
17–27	114.1	114.1/4 = 28.5
28–40	123.2	123.2/4 = 30.8

(d)

Section	SD	d.f.
1–16	8.09	15
17–27	9.69	10
28–40	8.38	12

Combined SD $= \sqrt{\{[\Sigma(\text{degrees of freedom}) (\text{within group SD})^2]/}$
$$[\Sigma \text{ degrees of freedom}]\}$$

$$= \sqrt{\{[15 \times (8.09)^2 + 10 \times (9.69)^2 + 12 \times (8.38)^2]/}$$
$$[15 + 10 + 12]\}$$

$$= 8.64$$

This is slightly higher than the localized SD of 7.83 owing to the high standard deviation in the second section. This section may possibly not be homogeneous but there are insufficient data to show a significant step change, the test statistic being marginally lower than the critical value. However, this type of situation will tend to inflate the estimate of the standard deviation and it is preferable to use the localized standard deviation as the measure of variability.

Chapter 10

Solution to Problem 1

(a) The plant manager believes reaction time is dependent upon concentration so reaction time is the dependent variable (y) and concentration is the independent variable (x).

(b) SD $(x) = 2.216$, SD $(y) = 34.902$, $\bar{x} = 11$, $\bar{y} = 340$, $\Sigma xy = 45510$.

$$\text{Cov} (x \text{ and } y) = \frac{\Sigma xy - n\bar{x}\bar{y}}{n-1}$$

$$= \frac{45510 - 12 \times 11 \times 340}{11}$$

$$= 57.272$$

(c) Correlation coefficient $(r) = \frac{\text{Cov} (xy)}{\text{SD} (x) \text{ SD} (y)}$

$$= \frac{57.272}{2.216 \times 34.902}$$

$$= 0.741$$

(d) Null hypothesis – Within the population of batches the correlation between reaction time and concentration is zero $(\rho = 0)$.

Alternative hypothesis – There is a positive correlation between reaction time and concentration $(\rho > 0)$.

Critical value – From Table H with a sample size of 12 and a one-sided test:

0.497 at a 5% significance level
0.658 at a 1% significance level.

Test statistic – The sample correlation coefficient of 0.741.

Decision – Reject the null hypothesis at the 1% significance level.

Conclusion – The reaction time is related to concentration in such a way that reaction time increases as concentration increases.

(e) $\text{Var}(x) = \{\text{SD}(x)\}^2$

$$= (2.216)^2$$

$$= 4.911$$

$$b = \frac{\text{Cov}(x \text{ and } y)}{\text{Var}(x)}$$

$$= \frac{57.272}{4.911}$$

$$= 11.66$$

$$a = \bar{y} - b\bar{x}$$

$$= 340 - 11.66 \times 11$$

$$= 211.7$$

The least squares regression equation is

$$y = 211.7 + 11.66x$$

(f) See Fig. S.14.

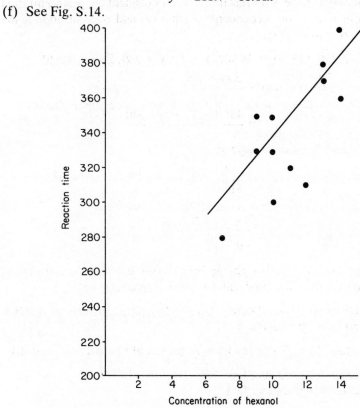

Figure S.14

(g) Percentage fit $= 100r^2$

$$= 100 \, (0.741)^2$$

$$= 54.9\%$$

(h) Residual sum of squares $= (n-1)(\text{SD of } y)^2 \, (1-r^2)$

$$= (12-1)(34.902)^2 \, (1-0.741^2) = 6042.2$$

Residual standard deviation $= \sqrt{(\text{residual sum of squares/} }$
$$\qquad\qquad \text{degrees of freedom})$$

$$= \sqrt{(6042.2/10)}$$

$$= 24.6$$

(i) Null hypothesis – There is no relationship between reaction time and concentration ($\beta = 0$).

Alternative hypothesis – There is a relationship between reaction time and concentration ($\beta \neq 0$).

Test statistic $= \sqrt{[(\text{new \% fit} - \text{old \% fit})(n-k-1)/(100 - \text{new \% fit})]}$

$$= \sqrt{[(54.9 - 0)(12 - 1 - 1)/(100 - 54.9)]}$$

$$= 3.49$$

Critical values – From the two-sided t-table with 10 degrees of freedom:

2.23 at the 5% significance level
3.17 at the 1% significance level.

Decision – We reject the null hypothesis at the 1% level of significance.

Conclusion – Reaction time is related to concentration.

(j) 95% confidence interval for the true intercept is given by:

$$a \pm t_{n-2}\,\text{RSD} \sqrt{\left(\frac{1}{n} + \frac{(\bar{x})^2}{(n-1)\,\text{Var}(x)}\right)}$$

$a = 211.7,\, t_{10} = 2.23,\, \text{RSD} = 24.6,\, n = 12,\, \bar{x} = 11,\, \text{Var}(x) = 4.911.$

$$211.7 \pm 2.23 \times 24.6 \sqrt{\left[\frac{1}{12} + \frac{11^2}{11 \times 4.911}\right]}$$

$$211.7 \pm 83.6$$

128.1 to 295.3

95% confidence interval for the slope is given by:

$$b \pm t_{n-2} \, \text{RSD} \sqrt{\left[\frac{1}{(n-1)\,\text{Var}\,(x)} \right]}$$

$$11.66 \pm 2.23 \times 24.6 \sqrt{\left[\frac{1}{11 \times 4.911} \right]}$$

$$11.66 \pm 7.46$$

4.20 to 19.12

(k) 95% confidence interval for the true mean reaction time at a specified concentration is given by:

$$(a + bX) \pm t_{n-2} \, \text{RSD} \sqrt{\left[\frac{1}{n} + \frac{(X-\bar{x})^2}{(n-1)\,\text{Var}\,(x)} \right]}$$

(i) When $X = 7$:

$$(211.7 + 11.66 \times 7) \pm 2.23 \times 24.6 \sqrt{\left[\frac{1}{12} + \frac{(7-11)^2}{11 \times 4.911} \right]}$$

$$293.3 \pm 33.8$$

259.5 to 327.1

(ii) When $X = 11$:

$$(211.7 + 11.66 \times 11) \pm 2.23 \times 24.6 \sqrt{\left[\frac{1}{12} + \frac{(11-11)^2}{11 \times 4.911} \right]}$$

$$340.0 \pm 15.8$$

324.2 to 355.8

(l) 95% confidence interval for a single reaction time at a specified concentration is given by:

$$a + bX \pm t_{n-2} \, \text{RSD} \sqrt{\left[1 + \frac{1}{n} + \frac{(X-\bar{x})^2}{(n-1)\,\text{Var}\,(x)} \right]}$$

(i) When $X = 7$:

$$(211.7 + 11.66 \times 7) \pm 2.23 \times 24.6 \sqrt{\left[1 + \frac{1}{12} + \frac{(7-11)^2}{11 \times 4.911} \right]}$$

$$293.3 \pm 64.4$$

228.9 to 357.7

(ii) When $X = 11$:

$$(211.7 + 11.66 \times 11) \pm 2.23 \times 24.6 \sqrt{\left[1 + \frac{1}{12} + \frac{(11-11)^2}{11 \times 4.911} \right]}$$

$$340.0 \pm 57.1$$
282.9 to 397.1

(m) In his search of past production records the plant manager will have unwittingly carried out many subjective significance tests and will have found out a set of data which looks 'significant'. The formal significance tests have just confirmed his informal judgement. This is therefore not in accordance with scientific method since both the theory and the conclusion have come from the same set of data.

Solution to Problem 2

(a)
$$\text{Cov}(xy) = (\Sigma xy - n\bar{x}\bar{y})/(n-1)$$
$$= (84689 - 6 \times 175 \times 80)/5$$
$$= 137.8$$
$$r = \frac{\text{Cov}(xy)}{\text{SD}(x)\,\text{SD}(y)}$$
$$= \frac{137.8}{18.708 \times \times 7.448}$$
$$= 0.989$$

(b)
$$b = \frac{\text{Cov}(xy)}{\text{Var}(x)}$$
$$= \frac{137.8}{(18.708)^2}$$
$$= 0.3937$$
$$a = \bar{y} - b\bar{x}$$
$$= 80 - 0.3937 \times 175$$
$$= 11.1$$

(c)
$$\text{RSD} = \sqrt{[(n-1)s_y^2(1-r^2)/(n-2)]}$$
$$= \sqrt{[5(7.448)^2(1-0.989^2)/4]}$$
$$= 1.24$$

95% confidence interval for the slope is given by:
$$b \pm t_{n-2}\,\text{RSD}\sqrt{\left[\frac{1}{(n-1)\text{Var}(x)}\right]}$$
$$0.394 \pm 2.78 \times 1.24 \sqrt{\left[\frac{1}{5 \times (18.708)^2}\right]}$$
$$0.394 \pm 0.082$$

(d) 95% confidence interval for the true value of y at a given value of x is:

$$(a + bx) \pm t_{n-2} \, \text{RSD} \sqrt{\left[\frac{1}{n} + \frac{(x - \bar{x})^2}{(n-1)\,\text{Var}(x)} \right]}$$

$$(11.1 + 0.3937x) \pm 2.78 \times 1.24 \sqrt{\left[\frac{1}{6} + \frac{(x - 175)^2}{5 \times (18.708)^2} \right]}$$

Wind-up speed	95% confidence for true value of birefringence
150	70.2 ± 2.49
160	74.1 ± 1.87
175	80.0 ± 1.41
190	85.9 ± 1.87
200	89.8 ± 2.49

For plots see Fig. S.15.

Summary

	Addy	Bolam	Cooper	Dawson
Covariance	137.8	101.0	298.0	21.28
Correlation coefficient	0.989	0.987	0.995	0.935
Intercept	11.1	9.3	10.5	13.5
Slope	0.394	0.404	0.397	0.380
Residual standard deviation	1.24	1.18	1.24	1.20
95% confidence interval for slope	±0.082	±0.093	±0.056	±0.199
95% CI for true line at:				
150	70.2 ± 2.5	69.9 ± 2.7	70.2 ± 2.0	70.5 ± 5.2
160	74.1 ± 1.9	73.9 ± 1.9	74.1 ± 1.6	74.3 ± 3.3
175	80.0 ± 1.4	80.0 ± 1.3	80.0 ± 1.4	80.0 ± 1.4
190	85.9 ± 1.9	86.1 ± 1.9	85.9 ± 1.6	85.7 ± 3.3
200	89.8 ± 2.5	90.1 ± 2.7	89.8 ± 2.0	89.5 ± 5.2

(e) No! Both the slope and the residual standard deviation are subject to sampling variation.

If we examine the confidence intervals for the slope coefficients we can see that these are wide compared with the range of the four experiments. They have been fortunate to obtain four such similar slopes.

The same can be said of the residual standard deviation. The confidence interval for a standard deviation can be found using Table 35 of Biometrica Tables for Statisticians (Pearson and Hartley, 1966). Use of this table, for Addy's experiment, as an example, gives values of 0.74 to 3.56 for a 95% confidence interval for the true population standard deviation.

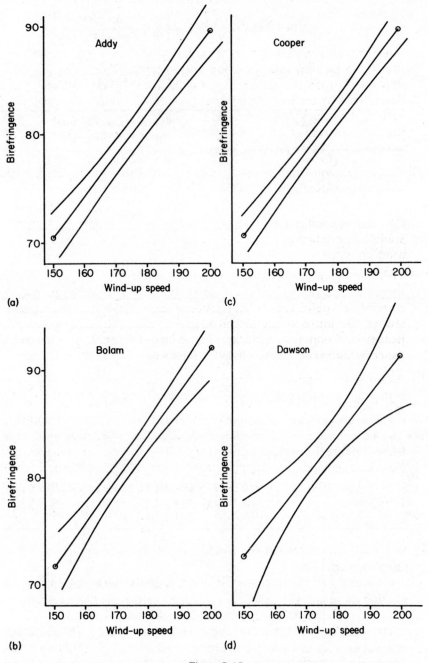

(a)

(c)

(b)

(d)

Figure S.15

(f) The confidence interval for the slope is given by:

$$t_{n-2}\,\mathrm{RSD}\sqrt{\left[\frac{1}{(n-1)\,\mathrm{Var}(x)}\right]}$$

Since the number of results were the same and there is little variability in the residual standard deviation, the confidence interval is mainly dependent upon the standard deviation of the x values. This value is determined by the experimental design. Dawson designed an experiment with all the values clustered round 175 and therefore s_x was very small. At the other extreme Cooper had three values at the two 'extremes' which maximizes s_x within the given range of wind-up speeds.

(g) The correlation coefficients together with the standard deviation of the wind-up speeds are given below:

	Dawson	Bolam	Addy	Cooper
Correlation coefficient	0.935	0.987	0.989	0.995
Standard deviation of wind-up speed	3.0	6.5	7.5	10.9

This table indicates a strong relationship between the range of wind-up speeds in the design, as indicated by the standard deviation, and the correlation coefficient. Interpretation of a correlation coefficient must always take into account the spread of the values. The designs chosen by Bolam and Cooper, with clusters of wind-up speeds, and large 'gaps', make the correlation coefficient difficult to interpret.

(h) Again this is dependent upon the design. Bolam has four central values with only two values at the 'extremes', while Cooper has no values in the centre.

(i) The confidence interval for Dawson's experiment is similar to the other three experiments within the range of wind-up speeds (165 to 185) used by him. Extrapolating outside this range produces high confidence intervals and will be little use in determining 'start-up' conditions for the new product.

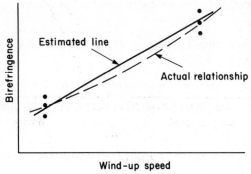

Figure S.16

(j) Yes! The danger is that the linearity assumptions may not be valid. With Addy's design we would obtain a strong indication of curvature because of the spread of points. With Cooper's design having only two values this is not possible. He could be markedly wrong as shown in Fig. S.16.

Chapter 11

Solution to Problem 1

(a) As the first independent variable (z) enters the equation the percentage fit becomes 50.7%. We test the statistical significance of the draw ratio as follows:

Null hypothesis – Tensile strength of the yarn is not related to the draw ratio (z).

Alternative hypothesis – Tensile strength of the yarn is related to the draw ratio (z).

$$\text{Test statistic} = \sqrt{[(\text{new \% fit} - \text{old \% fit})\,(n - k - 1)/(100 - \text{new \% fit})]}$$
$$= \sqrt{[(50.7 - 0)\,(12 - 1 - 1)/(100 - 50.7)]}$$
$$= 3.21$$

Critical values – From the two-sided t-table with 10 degrees of freedom:

2.23 at the 5% significance level
3.17 at the 1% significance level.

Decision – We reject the null hypothesis at the 1% level of significance.

Conclusion – We conclude that the tensile strength of the yarn is related to the draw ratio used in the drawing process.

As the second independent variable (x) enters the equation the percentage fit increases from 50.7% to 69.3%. We test the statistical significance of this increase as follows:

Null hypothesis – Tensile strength is not related to the spinning temperature (x).

Alternative hypothesis – Tensile strength is related to the spinning temperature (x).

$$\text{Test statistic} = \sqrt{[(\text{new \% fit} - \text{old \% fit})\,(n - k - 1)/(100 - \text{new \% fit})]}$$
$$= \sqrt{[(69.3 - 50.7)\,(12 - 2 - 1)/(100 - 69.3)]}$$
$$= 2.34$$

Critical values – From the two-sided t-test with 9 degrees of freedom:

2.26 at the 5% significance level
3.25 at the 1% significance level.

Decision – We reject the null hypothesis at the 5% level of significance.

Conclusion – We conclude the tensile strength of the yarn is related to the spinning temperature (x) as well as the draw ratio (z).

As the third independent variable (w) enters the equation the percentage fit increases from 69.3% to 73.8%. This small increase gives a test statistic equal to 1.17 which is less than the critical value for 5% significance (2.31, with 8 degrees of freedom). We therefore conclude that the tensile strength of the yarn is not related to the drying time (w).

(b) Before we draw any conclusions about the relationships between the independent variables on the one hand and the response on the other, we would be wise to examine the correlation matrix, paying particular attention to the intercorrelation of the three independent variables. Unfortunately the regression program did not print out a correlation matrix.

What can we deduce from the regression equations and their percentage fits? We will consider the equations in the order in which they were fitted:

(i) The percentage fit of the first equation is 50.7%. Dividing by 100 and taking the square root we obtain 0.712, so the correlation between draw ratio (z) and tensile strength (y) must be either +0.712 or −0.712. As the coefficient of z in the first equation is positive the correlation between y and z must be positive. We conclude therefore that $r_{zy} = 0.712$.

(ii) As the second independent variable (x) enters the equation the percentage fit increases from 50.7% to 69.3%. Comparing the first and second equations we note that *the coefficient of z does not change as x enters the equation*. We will conclude, therefore, that the correlation between x and z is equal to zero. (This conclusion is beyond dispute provided that there are no rounding errors in the coefficients.) It is quite possible that the research chemist chose the values of draw ratio (z) and spinning temperature (z) in his 12 trials so that the two variables would not be correlated with each other. Because there is zero correlation between the two independent variables in the second equation there is a simple relationship between the percentage fit and the correlations with the response:

$$\% \text{ fit} = 100(r_{zy}^2 + r_{xy}^2)$$
$$69.3 = 100(0.712^2 + r_{xy}^2)$$
$$r_{xy} = +0.431 \text{ or} -0.431$$

As the coefficient of x in the second equation is negative (and $r_{xy} = 0$) we can conclude that the correlation between the spin temperature (x) and tensile strength (y) is equal to −0.431.

(iii) As the third independent variable enters the equation there is a small increase in percentage fit which is not significant as we have demonstrated in part (a). We note that the coefficients of x and z change considerably as drying time (w) enters the equation. This indicates that r_{xw} and r_{zw} are rather high, but we are unable to say exactly what their values are. Nor are we able to estimate the correlation between drying time (w) and tensile strength. Our findings are summarized in the correlation matrix below:

	w	x	z	y
w	1.000	High	High	?
x	High	1.000	0.000	−0.431
z	High	0.000	1.000	0.712
y	?	−0.431	0.712	1.000

Solution to Problem 2

(1) response
(3) independent
(5) interaction
(7) response
(9) x
(11) z
(13) percentage fit
(15) degrees of freedom
(17) percentage fit
(19) covariance
(21) x
(23) residual sum of squares
(25) percentage fit
(27) residual standard deviation
(29) variation

(2) independent
(4) cross-product
(6) interaction
(8) value (or level)
(10) z
(12) t-test
(14) percentage fit
(16) x
(18) zero
(20) variance
(22) z
(24) residuals
(26) degrees of freedom
(28) response
(30) fixed

Solution to Problem 3

(a) Two of the six equations have tensile strength (y) as the dependent variable:

$$y = 1.100 + 0.0100z \qquad 85.0\% \text{ fit}$$
$$y = 2.867 + 0.0133x \qquad 53.3\% \text{ fit}$$

The percentage fit is calculated by squaring the appropriate correlation coefficient. Clearly the first of the two equations offers the better explanation of the variability in tensile strength. (This assertion is based solely on a *statistical* criterion, of course.)

(b) Percentage fit of the six equations are calculated from:

% fit $= 100\,r^2$

% fit $= 100(0.9220)^2 = 85.0\%$ for equation (1)

% fit $= 100(0.9220)^2 = 85.0\%$ for equation (2)

% fit $= 100(0.7303)^2 = 53.3\%$ for equation (3)

% fit $= 100(0.7303)^2 = 53.3\%$ for equation (4)

% fit $= 100(0.8416)^2 = 70.8\%$ for equation (5)

% fit $= 100(0.8416)^2 = 70.8\%$ for equation (6)

Note that equations (1) and (2) have the same percentage fit even though the two equations represent distinctly different lines on a graph. The same can be said of equations (3) and (4) or of equations (5) and (6).

(c) The partial correlation coefficient for y and x with z as the fixed variable can be calculated from:

$$[r_{xy} - r_{xz}r_{yz}]/\sqrt{[(1 - r_{xz}^2)\,(1 - r_{yz}^2)]}$$

$$= [0.7303 - (0.8416)\,(0.9220)]/\sqrt{[(1 - 0.8416^2)\,(1 - 0.9220^2)]}$$

$$= \frac{-0.04566}{0.20912} = -0.2183$$

Note that this partial correlation coefficient is negative though the simple correlation between y and x (0.7303) is positive. When the effect of the spinning temperature (z) has been taken into account it appears that an increase in drying time (x) will result in a *decrease* in tensile strength, rather than an *increase* as suggested by the positive simple correlation between x and y.

(d) Increase in % fit $= (100 - \text{old \% fit})\,(\text{partial correlation})^2$

$$= (100 - 85.0)\,(0.2183)^2$$

$$= 0.71\%$$

This minute increase in percentage fit may appear to be in conflict with the large correlation (0.7303) between x and y. The explanation lies in the huge correlation between the first independent variable (z) and the second independent variable (x).

(e) Percentage fit $= 100(r_{zy}^2 + r_{xy}^2 - 2r_{zy}r_{xy}r_{xz})/(1 - r_{xz}^2)$

$\qquad = 100[0.9220^2 + 0.7303^2 - 2(0.9220)(0.7303)$
$\qquad\qquad (0.8416)]/(1 - 0.8416^2)$

$\qquad = 100(0.8501 + 0.5333 - 1.1334)/(0.2917)$

$\qquad = 100(0.25)/(0.2917)$

$\qquad = 85.70\%$

This result is in agreement with our earlier finding, the percentage fit could be expected to increase by 0.71% from the base line of 85.00%.

(f) The missing lines are:

E   3.8   250   3.6   0.2
E   45   250   55   −10

(g) For the residuals in Table 11.6:

Mean $= 0.00$ Standard deviation $= 0.1225$

For the residuals in Table 11.7:

Mean $= 0.00$ Standard deviation $= 9.3541$

The covariance of the two sets of residuals is -0.2500.

The correlation of the two sets of residuals

$$= \frac{-0.2500}{(0.1225)\,(9.3541)}$$

$$= -0.2182$$

Note that this result is equal to the partial correlation coefficient calculated in (c).

Chapter 12

Solution to Problem 1

(a) Since we have three variables, *and* each variable has two levels, *and* each of the eight possible treatment combinations is used in one (and only one) trial the experiment is a 2^3 *factorial experiment.*

(b) You would be able to estimate:

(i) three main effects;
(ii) three two-variable interactions;
(iii) one three-variable interaction.

(c) Adding 6 to the response values will not alter the values of the estimates or the sums of squares.

(d)

	Temperature (°C)			
	80		90	
	pH		pH	
Speed of agitation	6	8	6	8
Slow	6	2	4	2
Fast	4	8	0	6

The table above is not the only possible three-way table. You may have produced a different arrangement and you may have called the variables A, B and C.

(e)

Speed	*Temperature (°C)*		*Speed*	*pH*		*Temperature*	*pH*	
	80	90		6	8		6	8
Slow	4	3	Slow	5	2	80	5	5
Fast	6	3	Fast	2	7	90	2	4

(f) Speed main effect $= 4.5 - 3.5 = 1.0$
Temperature main effect $= 3.0 - 5.0 = -2.0$
pH main effect $= 4.5 - 3.5 = 1.0$
Speed $\times$ Temp. interaction $= 3.5 - 4.5 = -1.0$
Speed $\times$ pH interaction $= 6.0 - 2.0 = 4.0$
Temp. $\times$ pH interaction $= 4.5 - 3.5 = 1.0$
Speed $\times$ temp. $\times$ pH interaction

$$= \frac{1}{2}\{[\frac{1}{2}(6+4) - \frac{1}{2}(2+0)] -$$
$$[\frac{1}{2}(6+8) - \frac{1}{2}(4+2)]\}$$
$$= 0.0$$

(g) The largest effect estimate is the speed $\times$ pH interaction. We will test this first:

Null hypothesis – There is no interaction between speed and pH.

Alternative hypothesis – There is an interaction between speed and pH.

$$\text{Test statistic} = \frac{\text{effect esimate}}{RSD/\sqrt{(p2^{n-2})}}$$

$$= \frac{4.0}{1.15/\sqrt{2}}$$

$$= 4.92$$

Critical values – From the two-sided t-table with 3 degrees of freedom:

   3.18 at the 5% significance level
   5.84 at the 1% significance level.

Decision – We reject the null hypothesis at the 5% level of significance.

Conclusion – We conclude that there *is* an interaction between speed of agitation and pH.

Further significance testing reveals that no other effect estimates are statistically significant.

Solution to Problem 2

(a)

Trial	s	t	p	$s \times t$	$s \times p$	$t \times p$	$s \times t \times p$
1	1	80	6	80	6	480	480
2	1	90	8	90	8	720	720
3	0	90	6	0	0	540	0
4	1	80	8	80	8	640	640
5	0	80	8	0	0	640	0
6	0	80	6	0	0	480	0
7	1	90	6	90	6	540	540
8	0	90	8	0	0	720	0

(b) If we examine the values of s and $(s \times t)$ all trials with $s = 0$ must also have $(s \times t) = 0$. All trials with $s = 1$ have $(s \times t) = 80$ or 90. Clearly this must result in a high correlation between s and $(s \times t)$. The use of 0, 1 for slow and fast speeds is the worst choice possible for obtaining high correlations between a main effect and its interactions.

(c) Using a coding of -1 and $+1$ for s and t gives:

Trial	s	t	$(s \times t)$
1	1	-1	-1
2	1	1	1
3	-1	1	-1
4	1	-1	-1
5	-1	-1	1
6	-1	-1	1
7	1	1	1
8	-1	1	-1

There is now a zero correlation between s and $(s \times t)$ compared with 0.99 with the uncoded data.

(d) Using the formula:

Test statistic $= \sqrt{[(\text{new \% fit} - \text{old \% fit}) (n - k - 1)/(100 - \text{new \% fit})]}$

% fit	Test statistic	Critical value 5%	Critical value 1%
16.67	1.09	2.45	3.71
28.43	0.91	2.57	4.03
77.08	2.91	2.78	4.60
94.60	3.12	3.18	5.84
98.77	2.60	4.30	9.92
100%	Infinite	12.71	63.66

(e) Clearly there are no variables or interactions above the critical value at the first stage and therefore the conclusion would be drawn that the variables and their interactions were not significant. Even with the full analysis as given in part (d), it is impossible to reach any conclusion apart from noting that the sudden jumps in percentage fit are a sign of a strange set of data.

(f) The final regression equation is:

$$y = 76.0 - 10s - 0.8t - 10p - 0.2(s \times t) + 4.0(s \times p) + 0.1(t \times p)$$

When $s = 1, t = 90, p = 8$:

$$y = 76.0 - (10 \times 1) - (0.8 \times 90) - (10 \times 8) - 0.2(1 \times 90)$$
$$+ 4(1 \times 8) + 0.1(90 \times 8)$$

$$= 76 - 10 - 72 - 80 - 18 + 32 + 72$$

$$= 0$$

(g) The three-variable interaction has an effect estimate of zero in Problem 1. Clearly all the variation can be explained without this interaction and it has no part to play in the multiple regression analysis.

Chapter 13

Solution to Problem 1

(a) There are four variables at two levels which would give 16 trials in a full experiment. This is therefore a ¼ (2^4) design.

(b)

Trial	A	B	C	D	AB	AC	AD	BC	BD	CD	ABC	ABD	ACD	BCD	ABCD
1	−	−	−	+	+	+	−	+	−	−	−	+	+	+	−
2	+	−	+	−	−	+	−	−	+	−	−	+	−	+	+
3	−	+	−	−	−	+	+	−	−	+	+	+	−	+	−
4	+	+	+	+	+	+	+	+	+	+	+	+	+	+	+

(c) The defining contrasts are:

   AC   ABD   BCD

(d) The alias groups are:

   (*A, C, BD, ABCD*)
   (*B, AD, CD, ABC*)
   (*D, AB, BC, ACD*)

Solution to Problem 2

(a) With the information you have been given in Chapter 13 the easiest way for you to answer this problem is to *find* a half replicate that will give the chemist the estimates he requires. Such a half replicate does exist. The design matrix for a 2^4 factorial experiment is given in Chapter 13. You might find it useful.

(b) Each of the design matrices below represents a half replicate of a 2^4 factorial design. Either of these half replicates will yield the estimates needed by the Chemist.

A	B	C	D		A	B	C	D
+	−	−	−		−	−	−	−
−	+	−	−		+	+	−	−
−	−	+	−		+	−	+	−
+	+	+	−		−	+	+	−
+	−	−	+		−	−	−	+
−	+	−	+		+	+	−	+
−	−	+	+		+	−	+	+
+	+	+	+		−	+	+	+

These two half replicates *together* constitute the full 2^4 experiment. For each of the half replicates the defining contrast is ABC and the alias pairs are (A, BC), (B, AC), (C, AB), $(D, ABCD)$, (AD, BCD), (BD, ACD), (CD, ABD).

Solution to Problem 3

(1) n
(2) 2
(3) 4
(4) 2
(5) interaction
(5) *t*-test
(7) residual standard deviation
(8) 8
(9) 4
(10) 8
(11) $y = a - bx + cz$
(12) $y = a + bx + cz + dxz$
(13) cross-product
(14) correlated
(15) scale (or standardize)
(16) zero
(17) replicate
(18) defining contrast
(19) alias pairs
(20) interaction XW
(21) interaction ZW
(22) main effect X
(23) interaction XZ
(24) interaction
(25) confounded

Chapter 14

Solution to Problem 1

(a) *Coding of design*
Use -1 for F1, low temperature, enzyme
and $+1$ for F4, high temperature, biozyme

Cell	Malt (A)	Temperature (B)	Yeast (C)
1	-1	-1	$+1$
2	-1	$+1$	-1
3	$+1$	-1	-1
4	$+1$	-1	$+1$
5	$+1$	-1	-1
6	-1	$+1$	-1
7	-1	$+1$	-1
8	$+1$	-1	$+1$
9	-1	-1	$+1$
10	-1	-1	$+1$

Correlation between malt (A) and temperature (B)
Using the formulae

$$r_{AB} = \frac{\text{covariance of } A \text{ and } B}{(\text{SD of } A)(\text{SD of } B)}$$

$$\text{Covariance} = \frac{\Sigma AB - n\bar{A}\bar{B}}{n-1}$$

$$\text{SD of } A = \sqrt{\left[\frac{\Sigma A^2 - n(\bar{A})^2}{n-1}\right]}$$

$$\text{SD of } B = \sqrt{\left[\frac{\Sigma B^2 - n(\bar{B})^2}{n-1}\right]}$$

Cell	A	B	A^2	B^2	AB
1	-1	-1	$+1$	$+1$	$+1$
2	-1	$+1$	$+1$	$+1$	-1
3	$+1$	-1	$+1$	$+1$	-1
4	$+1$	-1	$+1$	$+1$	-1
5	$+1$	-1	$+1$	$+1$	-1
6	-1	$+1$	$+1$	$+1$	-1
7	-1	$+1$	$+1$	$+1$	-1
8	$+1$	-1	$+1$	$+1$	-1
9	-1	-1	$+1$	$+1$	$+1$
10	-1	-1	$\cdot +1$	$+1$	$+1$
Σ	-2	-4	10	10	-4

$$\bar{A} = -2/10 = -0.2$$
$$\bar{B} = -4/10 = -0.4$$
$$\Sigma A^2 = 10$$
$$\Sigma B^2 = 10$$
$$\Sigma AB = -4$$

$$\text{Covariance} = \frac{-4 - 10(-0.2)(-0.4)}{9} = -0.53$$

$$\text{SD of } A = \sqrt{\left[\frac{10 - 10(-0.2)^2}{9}\right]} = \sqrt{1.07} = 1.03$$

$$\text{SD of } B = \sqrt{\left[\frac{10 - 10(-0.4)^2}{9}\right]} = 0.97$$

$$r_{AB} = \frac{\text{covariance of } A \text{ and } B}{(\text{SD of } A)(\text{SD of } B)}$$

$$= \frac{-0.53}{1.03 \times 0.97} = -0.53$$

Correlation matrix

	A	B	C
A	1.0	−0.53	0.0
B	−0.53	1.0	−0.65
C	0.0	−0.65	1.0

(b) The design can be represented as crosses at the corners of a cube with each cross denoting the experimental conditions for a particular cell (Fig. S.17). There are four corners without any crosses and it is desirable that two cells are chosen at one of these corners. It is also desirable that the imbalance in

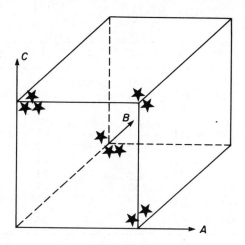

Figure S.17

the number of cells at different levels for A and B be rectified. However, our main concern must be to reduce the covariance of (AB) and (BC). Since both of these are negative it is imperative that the products of AB and AC are positive. Thus we can have A, B and C either all negative or all positive. All negative values will worsen the imbalance so all positive values are chosen. The extra cells are:

Cell	A	B	C
11	+1	+1	+1
12	+1	+1	+1

The experimental conditions for the new cells are

Cell	Malt	Temperature	Yeast
11	F4	High	Biozyme
12	F4	High	Biozyme

Note: These are the best two cells when only main effects are being investigated. The inclusion of two-factor interactions, which is essential for a realistic investigation, would lead to different cells.

(c) *Correlation of temperature (B) and malt (C)*

$$\Sigma B = -2 \qquad \bar{B} = -2/12 = -0.17$$
$$\Sigma C = +2 \qquad \bar{C} = 2/12 = +0.17$$
$$\Sigma B^2 = 12$$
$$\Sigma C^2 = 12$$
$$\Sigma BC = -4$$

$$\text{Covariance } B \text{ and } C = \frac{-4 - 12(-0.17)(0.17)}{11}$$

$$= -0.33$$

$$\text{SD of } B = \sqrt{\left(\frac{12 - 12(-0.17)^2}{11}\right)} = \sqrt{1.06} = 1.03$$

$$\text{SD of } C = \sqrt{\left(\frac{12 - 12(0.17)^2}{11}\right)} = \sqrt{1.06} = 1.03$$

$$\text{Correlation of } B \text{ and } C = \frac{\text{covariance of } B \text{ and } C}{(\text{SD of } B)(\text{SD of } C)}$$

$$= \frac{-0.33}{1.03 \times 1.03} = -0.31$$

Correlation matrix

	A	B	C
A	1.0	-0.17	0.17
B	-0.17	1.0	-0.31
C	0.17	-0.31	1.0

Solution to Problem 2

(a) *Coding of levels*

Temperature	270	-1
	275	0
	280	+1
Pressure	600	-1
	700	0
	800	+1
Concentration	0.15	-1
	0.20	0
	0.25	+1

The levels of quadratic and interactions are obtained by multiplying the appropriate levels from the main effects.

Design matrix

Cell	Temp.	Press	Conc.	$(Temp.)^2$	$(Press. \times conc.)$
1	0	0	−1	0	0
2	0	0	0	0	0
3	0	0	+1	0	0
4	0	−1	+1	0	−1
5	0	+1	+1	0	+1
6	−1	−1	+1	+1	−1
7	+1	−1	+1	+1	−1

(b) For ease of calculation:

Let temp. $= A$, press $= B$, conc. $= C$, (temp.)$^2 = D$, (press. $\times$ conc.) $= E$.

Correlation of B and E
Examination of the design matrix shows that B and E have identical values and therefore must have a correlation of 1.0.

Correlation of D and E

$$\Sigma\, D \; = \; 2 \qquad \bar{D} = \; 2/7 = \; 0.286$$
$$\Sigma\, E \; = -2 \qquad \bar{E} = -2/7 = -0.286$$
$$\Sigma\, D^2 \; = \; 2$$
$$\Sigma\, E^2 \; = \; 4$$
$$\Sigma\, DE = -2$$

Covariance of D and $E = \dfrac{2 - 7(0.286)\,(-0.286)}{6} = -0.238$

$$\text{SD of } D = \sqrt{\left[\frac{2 - 7(0.286)^2}{6}\right]} = \sqrt{0.238} = 0.487$$

$$\text{SD of } E = \sqrt{\left[\frac{4 - 7(-0.286))^2}{6}\right]} = \sqrt{0.571} = 0.755$$

Correlation of D and $E = \dfrac{-0.238}{0.487 \times 0.755} = -0.65$

Correlation matrix

	Temp.	Press.	Conc.	$(Temp.)^2$	$(Press. \times conc.)$
Temp.	1.0	0.0	0.0	0.0	0.0
Press.		1.0	−0.24	−0.65	1.0
Conc.			1.0	+0.37	−0.24
$(Temp.)^2$				1.0	−0.65
Press. $\times$ conc.					1.0

(c) The matrix reveals high correlations between three effects, the highest being press. and (press. × conc.) interaction. To reduce this correlation it is necessary that these effects have opposite signs. However, if press. has the same sign for both cells it will result in increasing the correlation between press. and conc. The choice of cells is given below.

Cell	Temp.	Press.	Conc.	$(Temp.)^2$	$(Press. \times conc.)$
8		+1	−1		−1
9		−1	−1		+1

Negative values of concentration also reduce the imbalance towards positive cells in the first seven cells.

It remains to choose values of temp. which reduce the negative correlation between $(temp.)^2$ and (press. × conc.) interaction. For cell 8, $(temp.)^2$ must take its lowest level of zero while for cell 9 it must have a level of +1. The level for temp. in cell 9 can therefore be either −1 or +1, it being irrelevant which is chosen. The full design for the extra two cells is given below.

Cell	Temp.	Press.	Conc.	$(Temp.)^2$	$(Press. \times conc.)$
8	0	+1	−1	0	−1
9	+1	−1	−1	+1	+1

(d) *Correlation between B and E*

$$\begin{aligned} \Sigma\, B &= -2 & \bar{B} &= -2/9 = -0.222 \\ \Sigma\, E &= -2 & \bar{E} &= -2/9 = -0.222 \\ \Sigma\, B^2 &= 6 \\ \Sigma\, E^2 &= 6 \\ \Sigma\, BE &= 2 \end{aligned}$$

$$\text{Covariance of } B \text{ and } E = \frac{2 - 9(-0.222)\,(-0.222)}{8} = 0.195$$

$$\text{SD of } B = \sqrt{\left(\frac{6 - 9(-0.222)^2}{8}\right)} = \sqrt{0.694} = 0.833$$

$$\text{SD of } E = \sqrt{\left(\frac{6 - 9(-0.222)^2}{8}\right)} = \sqrt{0.694} = 0.833$$

$$\text{Correlation of } B \text{ and } E = \frac{\text{covariance of } B \text{ and } E}{(\text{SD of } B)\,(\text{SD of } E)}$$

$$= \frac{0.195}{0.833 \times 0.833} = 0.28$$

Correlation matrix

	Temp.	Press.	Conc.	(Temp.)^2	(Press. $\times$ conc.)
Temp.	1.0	-0.19	-0.26	0.27	0.30
Press.		1.0	-0.24	-0.70	0.28
Conc.			1.0	0.08	-0.24
(Temp.)^2				1.0	-0.10
(Press. $\times$ conc.)					1.0

Thus it can be seen that two of the high correlations have been reduced but the correlation between (temp.)^2 and press. still remains high. It is impossible to reduce all three correlations with only two extra cells.

Chapter 15

Solution to Problem 1

(a) Total sum of squares

$$= (\text{overall SD})^2 (\text{degrees of freedom})$$

$$= (0.27775)^2 (14)$$

$$= 1.0800$$

(b) Within samples sum of squares

$$= \Sigma \{(\text{sample SD})^2 (\text{degrees of freedom})\}$$

$$= (0.100)^2 (4) + (0.18708)^2 (4) + (0.22361)^2 (4)$$

$$= 0.3800$$

(c) Between samples sum of squares

$$= (\text{SD of sample means})^2 (\text{number of samples} - 1)$$

$$\times (\text{number of observations on each sample})$$

$$= (0.26458)^2 (3 - 1) (5)$$

$$= 0.7000$$

(d)

Source of variation	Sum of squares	Degrees of freedom	Mean square
Between samples	0.7000	2	0.3500
Within samples	0.3800	12	0.0317
Total	1.0800	15	–

(e) Null hypothesis – There is no difference in impurity at the three depths in the tank.

Alternative hypothesis – There is variation in impurity from depth to depth.

$$\text{Test statistic} = \frac{\text{between samples mean square}}{\text{within samples mean square}}$$

$$= \frac{0.350}{0.0317}$$

$$= 11.04$$

Critical values – From the one-sided F-table with 2 and 12 degrees of freedom:

3.89 at the 5% significance level
6.93 at the 1% significance level.

Decision – We reject the null hypothesis at the 1% level of significance.

Conclusion – The concentration of impurity is different at different depths within the tank.

(f) Testing standard deviation $= \sqrt{}$(within samples mean square)

$$= \sqrt{0.0317}$$

$$= 0.178$$

(g) There is a strong indication in the data that the concentration of impurity is greater at greater depths, as we can see in the table below. The F-test has proved, beyond reasonable doubt, that the apparent increase in impurity is not simply due to testing error.

Depth (m)	0.5	2.5	4.5
Mean impurity	3.2	3.3	3.7

(h) In order to explore the relationship between impurity concentration and depth we could use regression analysis. We will do this in Problem 2.

Solution to Problem 2

(a) The total sum of squares is equal to 1.08. This was calculated in Problem 1. It is the same variation in impurity that we are exploring in both problems.

(b) Residual sum of squares

$$= (1 - r^2)(n - 1)(\text{SD of } y)^2$$

$$= (1 - 0.5787)(15 - 1)(0.27775)^2$$

$$= 0.455$$

(c) Regression sum of squares

$$= r^2 (n - 1) \, (\text{SD of } y)^2$$

$$= 0.5787 \, (15 - 1) \, (0.27775)^2$$

$$= 0.625$$

(d)

Source of variation	Sum of squares	Degrees of freedom	Mean square
Due to regression on x	0.6250	1	0.6250
Residual	0.4550	13	0.0350
Total	1.0800	14	

(e) Null hypothesis – Impurity concentration (y) is not related to depth (x).

Alternative hypothesis – Impurity concentration (y) is related to depth (x).

$$\text{Test statistic} = \frac{\text{regression mean square}}{\text{residual mean square}}$$

$$= \frac{0.6250}{0.0350}$$

$$= 17.86$$

Critical values – From the one-sided F-table with 1 and 13 degrees of freedom:

4.67 at the 5% significance level
9.07 at the 1% significance level.

Decision – We reject the null hypothesis at the 1% level of significance.

Conclusion – We conclude that the concentration of impurity is related to the depth from which the sample is taken.

We have explored the possibility that there is a *linear* relationship between impurity and depth. Perhaps the relationship is curved. With samples taken from three depths (0.5, 2.5 and 4.5 m) we could fit a quadratic curve, and we will do so in Problem 3.

Solution to Problem 3

(a) Percentage fit

$$= \frac{\text{due to regression on } X \text{ and } X^2 \text{ sum of squares}}{\text{total sum of squares}} \times 100$$

Due to regression on X and X^2 sum of squares

$$= (0.64815) \, (1.08)$$

$$= 0.7000$$

Residual sum of squares

= total sum of squares − regression sum of squares

= 1.0800 − 0.7000

= 0.3800

Source of variation	Sum of squares	Degrees of freedom	Mean square
Due to regression on X	0.6250	1	–
Due to introduction of X^2	0.0750	1	0.0750
Due to regression on X and X^2	0.7000	2	–
Residual	0.3800	12	0.0317
Total	1.0800	14	–

(b) Null hypothesis – The relationship between concentration of impurity (y) and depth (x) is a linear relationship.

Alternative hypothesis – The relationship between concentration of impurity (y) and depth (x) is a quadratic relationship.

$$\text{Test statistic} = \frac{\text{due to introduction of } X^2 \text{ mean square}}{\text{residual mean square}}$$

$$= \frac{0.0750}{0.0317}$$

$$= 2.37$$

Critical values – From the one-sided F-table with 1 and 12 degrees of freedom:

4.75 at the 5% significance level
9.33 at the 1% significance level.

Decision – We cannot reject the null hypothesis.

Conclusion – We conclude that the relationship between concentration of impurity (y) and depth (x) is a linear relationship.

(c) We see that the bottom three rows of the analysis of variance table in Problem 3 are exactly the same as the three rows in the table in Problem 1 except for the names given to the sources of variation.

Bibliography and further reading

Some of the books and papers listed below are referred to in the chapters, others are included for the benefit of readers who wish to study further.

Barnett, V. and Lewis, T. (1979) *Outliers in Statistical Data* Wiley.

Box, G. E. P., Hunter, W. G. and Hunter, J. S. (1978) *Statistics for Experimenters* Wiley.

British Standard 2846 *Guide to Statistical Interpretation of Data*
  Part 1 (1975) Routine analysis of quantitative data
  Part 2 (1981) Estimation of the mean: confidence interval
  Part 3 (1975) Determination of a statistical tolerance interval
  Part 4 (1976) Techniques of estimation and tests relating to means and variances
  Part 5 (1977) Power of test relating to means and variances
  Part 6 (1976) Comparison of two means in the case of paired observations

Caulcutt, R. and Boddy, R. (to be published 1983) *Statistics for Analytical Chemists* Chapman and Hall.

Chatfield, C. (1978) *Statistics for Technology* Chapman and Hall.

Chatterjee, S. and Price, B. (1977) *Regression Analysis by Example* Wiley.

Cox, D. R. (1958) *Planning of Experiments* Wiley.

Cox, D. R. and Snell, E. J. (1981) *Applied Statistics: Principles and Examples* Chapman and Hall.

Daniel, C. (1976) *Applications of Statistics to Industrial Experimentation* Wiley.

Daniel, C. and Wood, F. S. (1971) *Fitting Equations to Data* Wiley.

Davies, O. L. (1978) *Design and Analysis of Industrial Experiments* Longmans.

Davies, O. L. and Goldsmith, P. L. (1972) *Statistical Methods in Research and Production* Longmans.

Draper, N. and Smith, H. (1966) *Applied Regression Analysis* Wiley.

Johnson, N. and Leone, F. C. (1964) *Statistics and Experimental Design* Volume 1 and Volume 2, Wiley.

Moroney, M. J. (1966) *Facts from Figures* Penguin.

Pearson, E. S. and Hartley, H. O. (eds) (1966) *Biometrica Tables for Statisticians* Cambridge University Press.

Statistical tables

Table A   Normal distribution

z = standardized value
Prob. = probability of the standardized value being exceeded

z	Prob.	z	Prob.	z	Prob.	z	Prob.
0.00	0.5000	0.31	0.3783	0.62	0.2676	0.93	0.1762
0.01	0.4960	0.32	0.3745	0.63	0.2643	0.94	0.1736
0.02	0.4920	0.33	0.3707	0.64	0.2611	0.95	0.1711
0.03	0.4880	0.34	0.3669	0.65	0.2578	0.96	0.1685
0.04	0.4840	0.35	0.3632	0.66	0.2546	0.97	0.1660
0.05	0.4801	0.36	0.3594	0.67	0.2515	0.98	0.1635
0.06	0.4761	0.37	0.3557	0.68	0.2483	0.99	0.1611
0.07	0.4721	0.38	0.3520	0.69	0.2451	1.00	0.1587
0.08	0.4681	0.39	0.3483	0.70	0.2420	1.01	0.1562
0.09	0.4641	0.40	0.3446	0.71	0.2389	1.02	0.1539
0.10	0.4602	0.41	0.3409	0.72	0.2358	1.03	0.1515
0.11	0.4562	0.42	0.3372	0.73	0.2327	1.04	0.1492
0.12	0.4522	0.43	0.3336	0.74	0.2296	1.05	0.1469
0.13	0.4483	0.44	0.3300	0.75	0.2266	1.06	0.1446
0.14	0.4443	0.45	0.3264	0.76	0.2236	1.07	0.1423
0.15	0.4404	0.46	0.3228	0.77	0.2206	1.08	0.1401
0.16	0.4364	0.47	0.3192	0.78	0.2177	1.09	0.1379
0.17	0.4325	0.48	0.3156	0.79	0.2148	1.10	0.1357
0.18	0.4286	0.49	0.3121	0.80	0.2119	1.11	0.1335
0.19	0.4247	0.50	0.3085	0.81	0.2090	1.12	0.1314
0.20	0.4207	0.51	0.3050	0.82	0.2061	1.13	0.1292
0.21	0.4168	0.52	0.3015	0.83	0.2033	1.14	0.1271
0.22	0.4129	0.53	0.2981	0.84	0.2005	1.15	0.1251
0.23	0.4090	0.54	0.2946	0.85	0.1977	1.16	0.1230
0.24	0.4052	0.55	0.2912	0.86	0.1949	1.17	0.1210
0.25	0.4013	0.56	0.2877	0.87	0.1922	1.18	0.1190
0.26	0.3974	0.57	0.2843	0.88	0.1894	1.19	0.1170
0.27	0.3936	0.58	0.2810	0.89	0.1867	1.20	0.1151
0.28	0.3897	0.59	0.2776	0.90	0.1841	1.21	0.1131
0.29	0.3859	0.60	0.2743	0.91	0.1814	1.22	0.1112
0.30	0.3821	0.61	0.2709	0.92	0.1788	1.23	0.1093

z	Prob.	z	Prob.	z	Prob.	z	Prob.
1.24	0.1075	1.54	0.0618	1.84	0.0329	2.70	0.00346
1.25	0.1056	1.55	0.0606	1.85	0.0322	2.75	0.00297
1.26	0.1038	1.56	0.0594	1.86	0.0314	2.80	0.00255
1.27	0.1020	1.57	0.0582	1.87	0.0307	2.85	0.00219
1.28	0.1003	1.58	0.0571	1.88	0.0301	2.90	0.00187
1.29	0.0985	1.59	0.0559	1.89	0.0294	2.95	0.00159
1.30	0.0968	1.60	0.0548	1.90	0.0287	3.00	0.00135
1.31	0.0951	1.61	0.0537	1.91	0.0281	3.05	0.00114
1.32	0.0934	1.62	0.0526	1.92	0.0274	3.10	0.00097
1.33	0.0918	1.63	0.0516	1.93	0.0268	3.15	0.00082
1.34	0.0901	1.64	0.0505	1.94	0.0262	3.20	0.00069
1.35	0.0885	1.65	0.0495	1.95	0.0256	3.25	0.00058
1.36	0.0869	1.66	0.0485	1.96	0.0250	3.30	0.00048
1.37	0.0853	1.67	0.0475	1.97	0.0244	3.35	0.00040
1.38	0.0838	1.68	0.0465	1.98	0.0239	3.40	0.00034
1.39	0.0823	1.69	0.0455	1.99	0.0233	3.45	0.00028
1.40	0.0808	1.70	0.0446	2.00	0.0228	3.50	0.00023
1.41	0.0793	1.71	0.0436	2.05	0.0202	3.55	0.00019
1.42	0.0778	1.72	0.0427	2.10	0.0179	3.60	0.00016
1.43	0.0764	1.73	0.0418	2.15	0.0158	3.65	0.00013
1.44	0.0749	1.74	0.0409	2.20	0.0139	3.70	0.00011
1.45	0.0735	1.75	0.0401	2.25	0.0122	3.75	0.00009
1.46	0.0721	1.76	0.0392	2.30	0.0107	3.80	0.00007
1.47	0.0708	1.77	0.0384	2.35	0.0094	3.85	0.00006
1.48	0.0694	1.78	0.0375	2.40	0.0082	3.90	0.00005
1.49	0.0681	1.79	0.0367	2.45	0.0071	3.95	0.00004
1.50	0.0668	1.80	0.0359	2.50	0.0062	4.00	0.00003
1.51	0.0665	1.81	0.0351	2.55	0.0054		
1.52	0.0643	1.82	0.0344	2.60	0.0047		
1.53	0.0630	1.83	0.0336	2.65	0.0040		

Percentage points

	Significance level					

Two-sided			One-sided		
10% *(0.10)*	*5%* *(0.05)*	*1%* *(0.01)*	*10%* *(0.10)*	*5%* *(0.05)*	*1%* *(0.01)*
1.64	1.96	2.58	1.28	1.64	2.33

Table B   Critical values for the *t*-test

Degrees of freedom	Two-sided test 10% (0.10)	5% (0.05)	1% (0.01)	One-sided test 10% (0.10)	5% (0.05)	1% (0.01)
1	6.31	12.71	63.66	3.08	6.31	31.82
2	2.92	4.30	9.92	1.89	2.92	6.97
3	2.35	3.18	5.84	1.64	2.35	4.54
4	2.13	2.78	4.60	1.53	2.13	3.75
5	2.02	2.57	4.03	1.48	2.02	3.36
6	1.94	2.45	3.71	1.44	1.94	3.14
7	1.89	2.36	3.50	1.42	1.89	3.00
8	1.86	2.31	3.36	1.40	1.86	2.90
9	1.83	2.26	3.25	1.38	1.83	2.82
10	1.81	2.23	3.17	1.37	1.81	2.76
11	1.80	2.20	3.11	1.36	1.80	2.72
12	1.78	2.18	3.06	1.36	1.78	2.68
13	1.77	2.16	3.01	1.35	1.77	2.65
14	1.76	2.15	2.98	1.35	1.76	2.62
15	1.75	2.13	2.95	1.34	1.75	2.60
16	1.75	2.12	2.92	1.34	1.75	2.58
17	1.74	2.11	2.90	1.33	1.74	2.57
18	1.73	2.10	2.88	1.33	1.73	2.55
19	1.73	2.09	2.86	1.33	1.73	2.54
20	1.72	2.08	2.85	1.32	1.72	2.53
25	1.71	2.06	2.78	1.32	1.71	2.49
30	1.70	2.04	2.75	1.31	1.70	2.46
40	1.68	2.02	2.70	1.30	1.68	2.42
60	1.67	2.00	2.66	1.30	1.67	2.39
120	1.66	1.98	2.62	1.29	1.66	2.36
Infinite	1.64	1.96	2.58	1.28	1.64	2.33

Table C Critical values for the F-test

One-sided at 5% significance level

Degrees of freedom for smaller variance	Degrees of freedom for larger variance														
	1	2	3	4	5	6	7	8	9	10	12	15	20	60	Infinity
1	161.4	199.5	215.7	224.6	230.2	234.0	236.8	238.9	240.5	241.9	243.9	246.0	248.0	252.2	254.3
2	18.51	19.00	19.16	19.25	19.30	19.33	19.35	19.37	19.38	19.40	19.41	19.43	19.45	19.48	19.50
3	10.13	9.55	9.28	9.12	9.01	8.94	8.89	8.85	8.81	8.79	8.74	8.70	8.66	8.57	8.53
4	7.71	6.94	6.59	6.39	6.26	6.16	6.09	6.04	6.00	5.96	5.91	5.86	5.80	5.69	5.63
5	6.61	5.79	5.41	5.19	5.05	4.95	4.88	4.82	4.77	4.74	4.68	4.62	4.56	4.43	4.36
6	5.99	5.14	4.76	4.53	4.39	4.28	4.21	4.15	4.10	4.06	4.00	3.94	3.87	3.74	3.67
7	5.59	4.74	4.35	4.12	3.97	3.87	3.79	3.73	3.68	3.64	3.57	3.51	3.44	3.30	3.23
8	5.32	4.46	4.07	3.84	3.69	3.58	3.50	3.44	3.39	3.35	3.28	3.22	3.15	3.01	2.93
9	5.12	4.26	3.86	3.63	3.48	3.37	3.29	3.23	3.18	3.14	3.07	3.01	2.94	2.79	2.71
10	4.96	4.10	3.71	3.48	3.33	3.22	3.14	3.07	3.02	2.98	2.91	2.85	2.77	2.62	2.54
12	4.75	3.89	3.49	3.26	3.11	3.00	2.91	2.85	2.80	2.75	2.69	2.62	2.54	2.38	2.30
15	4.54	3.68	3.29	3.06	2.90	2.79	2.71	2.64	2.59	2.54	2.48	2.40	2.33	2.16	2.07
20	4.32	3.49	3.10	2.87	2.71	2.60	2.49	2.45	2.39	2.35	2.28	2.20	2.12	1.95	1.84
60	4.00	3.15	2.76	2.53	2.37	2.25	2.17	2.10	2.04	1.99	1.92	1.84	1.75	1.53	1.39
Infinity	3.84	3.00	2.60	2.37	2.21	2.10	2.01	1.94	1.88	1.83	1.75	1.67	1.57	1.32	1.00

One-sided at 1% significance level

Degrees of freedom for smaller variance	Degrees of freedom for larger variance														
	1	2	3	4	5	6	7	8	9	10	12	15	20	60	Infinity
1	4052	5000	5403	5625	5764	5859	5928	5982	6022	6056	6106	6157	6209	6313	6366
2	98.50	99.00	99.17	99.25	99.30	99.33	99.36	99.37	99.39	99.40	99.42	99.43	99.45	99.48	99.50
3	34.12	30.82	29.46	28.71	28.24	27.91	27.67	27.49	27.35	27.23	27.05	26.87	26.69	26.32	26.13
4	21.20	18.00	16.69	15.98	15.52	15.21	14.98	14.80	14.66	14.55	14.37	14.20	14.02	13.65	13.46
5	16.26	13.27	12.06	11.39	10.97	10.67	10.46	10.29	10.16	10.05	9.89	9.72	9.55	9.20	9.02
6	13.75	10.92	9.78	9.15	8.75	8.47	8.26	8.10	7.98	7.87	7.72	7.56	7.40	7.06	6.88
7	12.25	9.55	8.45	7.85	7.46	7.19	6.99	6.84	6.72	6.62	6.47	6.31	6.16	5.82	5.65
8	11.26	8.65	7.59	7.01	6.63	6.37	6.18	6.03	5.91	5.81	5.67	5.52	5.36	5.03	4.86
9	10.56	8.02	6.99	6.42	6.06	5.80	5.61	5.47	5.35	5.26	5.11	4.96	4.81	4.48	4.31
10	10.04	7.56	6.55	5.99	5.64	5.39	5.20	5.06	4.94	4.85	4.71	4.56	4.41	4.08	3.91
12	9.33	6.93	5.95	5.41	5.06	4.82	4.64	4.50	4.39	4.30	4.16	4.01	3.86	3.54	3.36
15	8.68	6.36	5.42	4.89	4.56	4.32	4.14	4.00	3.89	3.80	3.67	3.52	3.37	3.05	2.87
20	8.10	5.85	4.94	4.43	4.10	3.87	3.70	3.56	3.46	3.37	3.23	3.09	2.94	2.61	2.42
60	7.08	4.98	4.13	3.65	3.34	3.12	2.95	2.82	2.72	2.63	2.50	2.35	2.20	1.84	1.60
Infinity	6.63	4.61	3.78	3.32	3.02	2.80	2.64	2.51	2.41	2.32	2.18	2.04	1.88	1.47	1.00

Two-sided at 5% significance level

Degrees of freedom for smaller variance	Degrees of freedom for larger variance														
	1	2	3	4	5	6	7	8	9	10	12	15	20	60	Infinity
1	647.8	799.5	864.2	899.6	921.8	937.1	948.2	956.7	963.3	968.6	976.7	984.9	993.1	1010.0	1061.8
2	38.51	39.00	39.17	39.25	39.30	39.33	39.36	39.37	39.39	39.40	39.41	39.43	39.45	39.48	39.50
3	17.44	16.04	15.44	15.10	14.88	14.73	14.62	14.54	14.47	14.42	14.34	14.25	14.17	13.99	13.90
4	12.22	10.65	9.98	9.60	9.36	9.20	9.07	8.98	8.90	8.84	8.75	8.66	8.56	8.36	8.26
5	10.01	8.43	7.76	7.39	7.15	6.98	6.85	6.76	6.68	6.62	6.52	6.43	6.33	6.12	6.02
6	8.81	7.26	6.60	6.23	5.99	5.82	5.70	5.60	5.52	5.46	5.37	5.27	5.17	4.96	4.85
7	8.07	6.54	5.89	5.52	5.29	5.12	4.99	4.90	4.82	4.76	4.67	4.57	4.47	4.25	4.14
8	7.57	6.06	5.42	5.05	4.82	4.65	4.53	4.43	4.36	4.30	4.20	4.10	4.00	3.78	3.67
9	7.21	5.71	5.08	4.72	4.48	4.32	4.20	4.10	4.03	3.96	3.87	3.77	3.67	3.45	3.33
10	6.94	5.46	4.83	4.47	4.24	4.07	3.95	3.85	3.78	3.72	3.62	3.52	3.42	3.20	3.08
12	6.55	5.10	4.47	4.12	3.89	3.73	3.61	3.51	3.44	3.37	3.28	3.18	3.07	2.85	2.72
15	6.20	4.77	4.15	3.80	3.58	3.41	3.29	3.20	3.12	3.06	2.96	2.86	2.76	2.52	2.40
20	5.87	4.46	3.86	3.51	3.29	3.13	3.01	2.91	2.84	2.77	2.68	2.57	2.46	2.22	2.09
60	5.29	3.93	3.34	3.01	2.79	2.63	2.51	2.41	2.33	2.27	2.17	2.06	1.94	1.67	1.48
Infinity	5.02	3.69	3.12	2.79	2.57	2.41	2.29	2.19	2.11	2.05	1.94	1.83	1.71	1.39	1.00

Two-sided at 1% significance level

Degrees of freedom for smaller variance	Degrees of freedom for larger variance														
	1	2	3	4	5	6	7	8	9	10	12	15	20	60	Infinity
1	16211	20000	21615	22500	23056	23437	23715	23925	24091	24224	24426	24630	24836	25253	25465
2	198.5	199.0	199.2	199.2	199.3	199.3	199.4	199.4	199.4	199.4	199.4	199.4	199.4	199.5	199.5
3	55.55	49.80	47.47	46.19	45.39	44.84	44.43	44.13	43.88	43.69	43.29	43.08	42.78	42.15	41.83
4	31.33	26.28	24.26	23.15	22.46	21.97	21.62	21.35	21.14	20.97	20.70	20.04	20.17	19.61	19.32
5	22.78	18.31	16.53	15.56	14.94	14.51	14.20	13.96	13.77	13.62	13.38	13.15	12.90	12.40	12.14
6	18.63	14.54	12.92	12.03	11.46	11.07	10.79	10.57	10.39	10.25	10.03	9.81	9.59	9.12	8.88
7	16.24	12.40	10.88	10.05	9.52	9.16	8.89	8.68	8.51	8.38	8.18	7.97	7.75	7.31	7.08
8	14.69	11.04	9.60	8.81	8.30	7.95	7.69	7.50	7.34	7.21	7.01	6.81	6.61	6.18	5.95
9	13.61	10.11	8.72	7.96	7.47	7.13	6.88	6.69	6.54	6.42	6.23	6.03	5.83	5.41	5.19
10	12.83	9.43	8.08	7.34	6.87	6.54	6.30	6.12	5.97	5.85	5.66	5.47	5.27	4.86	4.64
12	11.75	8.51	7.23	6.52	6.07	5.76	5.52	5.35	5.20	5.09	4.91	4.72	4.53	4.12	3.90
15	10.80	7.70	6.48	5.80	5.37	5.07	4.85	4.67	4.54	4.42	4.25	4.07	3.88	3.48	3.26
20	9.94	6.99	5.82	5.17	4.76	4.47	4.26	4.09	3.96	3.85	3.68	3.50	3.32	2.92	2.69
60	8.49	5.79	4.73	4.14	3.76	3.49	3.29	3.13	3.01	2.90	2.74	2.57	2.39	1.96	1.69
Infinity	7.88	5.30	4.28	3.72	3.35	3.09	2.90	2.74	2.62	2.52	2.36	2.19	2.00	1.53	1.00

Table D Critical values for the triangular test

Number of sets of three samples	Significance level		
	5% (0.05)	1% (0.01)	0.1% (0.001)
5	3½	4½	–
6	4½	5½	–
7	4½	5½	6½
8	5½	6½	7½
9	5½	6½	7½
10	6½	7½	8½
11	6½	7½	9½
12	7½	8½	9½
13	7½	8½	10½
14	8½	9½	10½
15	8½	9½	11½
16	8½	10½	11½
17	9½	10½	12½
18	9½	11½	12½
19	10½	11½	13½
20	10½	12½	13½
25	12½	14½	16½
30	14½	16½	18½
35	16½	18½	21½
40	18½	20½	23½
45	20½	23½	25½
50	22½	25½	27½
60	26½	29½	32½
70	30½	33½	36½
80	34½	37½	40½
90	37½	41½	44½
100	41½	45½	48½

Table E Critical values for the chi-squared test

Degrees of freedom	Significance level	
	5% (0.05)	5% (0.01)
1	3.841	6.635
2	5.991	9.210
3	7.816	11.35
4	9.488	13.28
5	11.07	15.08
6	12.59	16.81
7	14.07	18.49
8	15.51	20.09
9	16.92	21.67
10	18.31	23.21
11	19.68	24.72
12	21.03	26.22
13	22.36	27.69
14	23.68	29.14
15	25.00	30.58
16	26.30	32.00
17	27.59	33.41
18	28.87	34.81
19	30.14	36.19
20	31.41	37.57

Table F Confidence interval for a population standard deviation

| Degrees of freedom | Level of confidence | | | |
| | 90% | | 95% | |
	L_1	L_2	L_1	L_2
1	0.51	15.9	0.45	31.9
2	0.58	4.42	0.52	6.28
3	0.62	2.92	0.57	3.73
4	0.65	2.37	0.60	2.87
5	0.67	2.09	0.62	2.45
6	0.69	1.92	0.64	2.20
7	0.71	1.80	0.66	2.04
8	0.72	1.71	0.68	1.92
9	0.73	1.65	0.69	1.83
10	0.74	1.59	0.70	1.75
12	0.76	1.52	0.72	1.65
15	0.77	1.44	0.74	1.55
20	0.80	1.36	0.77	1.44
24	0.81	1.32	0.78	1.39
30	0.83	1.27	0.80	1.34
40	0.85	1.23	0.82	1.28
60	0.87	1.18	0.85	1.22
Infinity	1.00	1.00	1.00	1.00

When estimating a population standard deviation (σ) by means of a sample standard deviation (s) a confidence interval for σ is given by:

$$\text{Lower limit} = L_1 s$$

$$\text{Upper limit} = L_2 s$$

Table G   Critical values for Dixon's test

Sample size	Critical values 5%	Critical values 1%
3	0.970	0.994
4	0.829	0.926
5	0.710	0.821
6	0.628	0.740
7	0.569	0.680
8	0.608	0.717
9	0.564	0.672
10	0.530	0.635
11	0.502	0.605
12	0.479	0.579
13	0.611	0.697
14	0.586	0.670
15	0.565	0.647
16	0.546	0.627
17	0.529	0.610
18	0.514	0.594
19	0.501	0.580
20	0.489	0.567
21	0.478	0.555
22	0.468	0.544
23	0.459	0.535
24	0.451	0.526
25	0.443	0.517
26	0.436	0.510
27	0.429	0.502
28	0.423	0.495
29	0.417	0.489
30	0.412	0.483
31	0.407	0.477
32	0.402	0.472
33	0.397	0.467
34	0.393	0.462
35	0.388	0.458
36	0.384	0.454
37	0.381	0.450
38	0.377	0.446
39	0.374	0.442
40	0.371	0.438

Table H   Critical values of the product moment correlation

Degrees of freedom	Two-sided test		One-sided test	
	5% (0.05)	1% (0.01)	5% (0.05)	1% (0.01)
2	0.950	0.990	0.900	0.980
3	0.878	0.959	0.805	0.934
4	0.811	0.917	0.729	0.882
5	0.754	0.875	0.669	0.833
6	0.707	0.834	0.621	0.789
7	0.666	0.798	0.582	0.750
8	0.632	0.765	0.549	0.715
9	0.602	0.735	0.521	0.685
10	0.576	0.708	0.497	0.658
11	0.553	0.684	0.476	0.634
12	0.532	0.661	0.457	0.612
13	0.514	0.641	0.441	0.592
14	0.497	0.623	0.426	0.574
15	0.482	0.606	0.412	0.558
20	0.423	0.537	0.360	0.492
30	0.349	0.449	0.296	0.409
40	0.304	0.393	0.257	0.358
60	0.250	0.325	0.211	0.295

Table I   Critical values for the cusum test

Length of span	Significance level	
	5% (0.05)	1% (0.01)
5	2.5	3.3
6	2.7	3.6
7	2.9	4.0
8	3.1	4.3
9	3.3	4.6
10	3.6	4.9
12	4.0	5.3
15	4.5	5.8
20	5.3	6.6
25	6.0	7.3
30	6.7	8.0
40	8.0	9.3
50	9.1	10.4
60	10.0	11.3
70	10.8	12.2
80	11.5	12.9
90	12.2	13.7
100	12.8	14.3

Index